Human Growth and Nutrition: A Biocultural Synthesis

Human Growth and Nutrition: A Biocultural Synthesis

Rajesh K. Gautam
and
Prasanna K. Patra

KALPAZ

ISBN: 978-93-5128-204-4

Price: ₹ 990

First Published, 2016

Published by

Kalpaz Publications
C-30, Satyawati Nagar,
Delhi – 110052
E-mail: kalpaz@hotmail.com
Ph.: 9212142040

Printed at: G. Print Process, Delhi

Cataloging in Publication Data—DK
Courtesy: D.K. Agencies (P) Ltd. <docinfo@dkagencies.com>

IUAES-Intercongress (5th : 2012 : Kalinga Institute of Industrial Technology, Bhubaneswar)
Human growth and nutrition : a biocultural synthesis / [editors], Rajesh K. Gautam and Prasanna K. Patra.
pages cm
Collection of papers of a special panel as part of 5th IUAES International Conference held on 26-30 November 2012 at Kalinga Institute of Industrial Technology (KIIT), Bhubaneswar.
Includes bibliographical references.
ISBN 9789351282044

1. Medical anthropology—India—Congresses. 2. Human growth—India—Congresses. 3. Nutrition—India—Congresses. I. Gautam, Rajesh K., 1970- editor. II. Patra, Prasanna Kumar, editor. III. Kalinga Institute of Industrial Technology, Bhubaneswar.host institution. IV. Title.

GN296.5.I4I93 2012 DDC 306.4610954 23

Contents

Theme – 2
Under-Nutrition and Disease

Theme – 3
Adult Nutrition, BMI and Puberty

Theme – 4
Physical activity and body composition

Theme – 5
Nutritional status and associated factors

Foreword

The present volume prepares Human scientists or anthropologists for the exciting and challenging area of human growth and development which assists in understanding its notion in an international arena as we move towards the twenty first century.

The volume edited by Dr. Rajesh Gautam and Dr. P. K. Patra who have received training from Japan, U.K., Netherland and U.S.A. have focused on Indian perspectives of human growth and development which is the crucial need of the hour in India. Each paper in the present volume speaks the variegated areas of the research, innovative ideas and advanced techniques in the field of growth and development which can be implemented in the management of research focusing on human variations, adaptations and morpho-physiological dynamics.

The status of health and development of a population or a community is reflected in the growth of its children. The pattern of that growth is shaped by the dynamic correlation between biological and cultural phenomena. This volume provides fascinating reading with diverse and unique examples of bio-social perspective into growth and nutrition based on empirical evidences from varied bio-physical, socio-cultural and demographic backgrounds.

The highlights of this volume lie on several factors that include the wider coverage to topics related to and factors that shape human growth and nutrition. It includes contributions that cover wider geographical, ecological and regional variations.

I am sure; this book should be of interest to those working in the fields of human growth and development. There is information about pre-natal and post natal growth of human child as well as growth from infancy to adulthood.

There are 16 chapters divided into 6 different themes, having comprehensive information on growth pattern form early embryonic stage to adulthood, under nutrition, puberty, body mass index, physical activity and body composition, nutritional status and associated factors and the last theme is food security and health care.

There is information on growth pattern of children belonging to different ethnic origin, socio-economic background and genetic makeup viz. Sicklers, Tibetan, Tribal and so on. Probably, there is first reporting on growth of homozygous Sicklers.

I found the book very useful for biological anthropologist, medical students, nutritionists and other walk of scholars like- geography, economics, health sciences, sociology etc. Beside that the book would also be useful for administrators, policy makers, planners, industrialists (furniture, garments, toys) etc.

Dated: 20th January, 2016 **Professor Dr. A. K. Kapoor**
University of Delhi
Delhi

Acknowledgements

This collection of papers in the form of an edited volume entitled '*Human Growth and Nutrition: A Biocultural Synthesis*' is the result of collaborative efforts of a group of scholars – established and budding-studying the varied aspects of human growth and development. It takes into account diverse social, demographic and policy contexts that shape the dynamism of growth and nutrition among rural and tribal groups in different parts of India. This volume is the culmination of a special panel organised by both the editors as part of Fifth IUAES international conference held on 26-30 November 2012 at Kalinga Institute of Industrial Technology (KIIT), Bhubaneswar.

The editors would like to thank all the contributing authors for their support to the volume by submitting their manuscripts in time, for being exceptionally tolerant about the shifting deadlines to see the volume being published. The editors would like to thank Professor Deepak K. Behera, the Organising Secretary of Fifth IUAES for kindly giving us the opportunity to hold a panel on the theme 'Human Growth and Nutrition: A Biocultural Synthesis' that transpired us to bring out this volume. Finally, without the enormous support of Mr Amit Garg of Kalpaz Publications for bringing out the book in record time, this could not be possible.

Lastly but not least, the authors would like to thank Professor A.K. Kapoor, a subject expert and a teacher par excellence for kindly writing the Forward to the book.

—Rajesh K. Gautam and
Prasanna K. Patra

Preface

This volume '*Human Growth and Nutrition: A Biocultural Synthesis*' is a synthesis of many strands of evidence for a biocultural model of human growth and nutrition. Synthesis of biocultural approaches recognises the pervasiveness and dynamism of interactions between biological and cultural phenomena, and they explicitly strive to integrate biological, sociocultural, environmental, and other kinds of data. This biocultural model allows us to better understand the causes of variation, plasticity in human growth and development, and nutrition. It helps us to identify and improve factors that impair health, growth, and development. The biocultural perspective focuses on the recurring interaction between the biology of human growth and development and the sociocultural environment. Not only do the latter influence the former, but human developmental biology modifies social and cultural processes as well.

Though the foundation of anthropology is based on a biocultural perspective, there are not many topics that are more rooted in both the biological and socio-cultural aspects of humankind. We need food and water to survive, but what we eat, how we prepare it, who we consume it with and how provisions for availability of such foods are made are influenced by sociocultural, economic and political factors. Human growth and development and nutritional intake ultimately determines individual and social health. This volume, a collection of very different aspects of nutrition and health, takes into account diverse social, demographic and policy contexts that shape the dynamism of growth and nutrition among rural and tribal groups in different parts of India. The following is a brief overview of the volume which is divided into six different yet inter-connected themes and each theme consists of one or more research paper/s contributed by scholars – both established in their fields and are budding.

There are four chapters under first theme: 'Human Growth and Develpment' which deal with human growth and development from its very inception till adulthood among contemporary Indian population. The first chapter entitled 'Ultrasonographic Prenatal Growth Study of Human Foetuses among the Meitei of Manipur' by T. Ursa and S. Jibonkumar Singh discussed in detail about the secular nature of prenatal growth among the Meitei Foetuses of Manipur and suggested that factors behind may be varied. It highlighted that the reason for the progressive secular growth inside utero had been taken as due to genetic, change in maternal lifestyle and nutrition and better access to antenatal care programmes. The study recommends that the present findings of the foetal biometric growth parameters needs to be employed as a reference value of foetal growth chart of the population as there is a need to accumulate a basic biometric data for every biologically varying populations.

The second chapter entitled 'Growth Pattern of stature and body weight during childhood and adolescence: A cross-sectional study among School Boys of Guwahati, Assam, India' carried out by Baruah *et al.*, has displayed the growth pattern of human child during its school age. They have concluded that though the Guwahati school boys are heavier and taller than our national average they are lighter and shorter than NCHS reference.

The third chapter is on 'Pattern of Growth among Homozygous Sicklers: Cross-sectional Study among Mehra a Scheduled Caste of Mandla District, Madhya Pradesh' by J. Jhariya *et al.*, highlight that the homozygous children for sickle cell gene have delayed growth and development at all ages specially after five years of age. Height for age is 15 per cent delayed as compared to normal children whereas body weight for age is 46 per cent delayed.

The fourth chapter is contributed by K.C. Satapathy which highlights the growth and nutritional status of adolescent Tibetans born and brought up in coastal Odisha and he has compared that data with high altitude studies by using standard anthropometric technique. His study findings emphasise that the Tibetans in costal Odisha show a trend towards becoming taller and heavier when compared in terms of their height, weight, BMI, and birth weight. Also, the blood pressure level of adolescent Tibetans show a trend towards higher prevalence of hypertension and is similar to that of high altitude studies.

The fifth and last chapter under the first theme entiled 'Human Growth and development was contributed by Chittaranjan Mandal, the chapter entiled 'Growth pattern of the Adolescent: A Cross-sectional study among the Bhotia of Uttaranchal, India'. In his chapter detailed comparative growth pattern of human child during adolescence is discussed with special reference to Bhotias and other ethnic groups from central Himalayan regions. The chapter has illustrated that the weight and height of Bhotia boys are below the NCHS standards at all ages. Bhotia boys invariably have more weight and height than ICMR standards at all age groups. Based on the data of children upto 14 years of age, the Author had concluded that Garhwali Brahmins are heavier than the Bhotia, thereafter, the Bhotia outweigh the Garhwali Brahmins. So far height is concerned, the Garhwali Brahmins are taller than Bhotias at all age groups.

Second theme is focused on 'Undernutrition and Disease'. There are three chapters (6, 7 and 8) under this theme. Chapter 6 is contributed by Mandal *et al.*, on the 'prevalence of thinness, overweight and obesity among Karbi tribal children aged 5-12 years of Karbi Anglong, Assam, Northeast India. The chapter is focused on the assessment of thinness, overweight and obesity using newly proposed cut-offs of Cole *et al.*, (2000; 2007). The *Comparisons with other studies hint that* the differences in the prevalence of under nutrition in population *could be because of* genetic and ethnic makeup *that* could play significant roles in such prevalence *of under nutrition*. There were sex specific differences in the prevalence of *under nutrition* between boys and girls, but girls were observed to be more affected by under nutrition than boys in most cases.

Chapter 7 entitled 'Co-prevalence of high rate of undernutrition and hypertension among six tribal population groups of India' is contributed by Gautam K Kshatriya and Subhendu K. Acharya on 'Co-prevalence of high rate of under-nutrition and hypertension among six tribal population groups of India', highlights the findings of trends in height and weight among 40-50 years and 50-60 years age groups in males and all the four age groups in females. It shows continued and constant poor health. The finding with respect to prevalence of hypertensive SBP and hypertensive DBP shows that hypertensive SBP risk is higher among females than males of elderly age groups, while such risk with respect to hypertensive DBP is higher among females than males in all the age groups.

Chapter 8 entitled 'Prevalence of Anemia among Tribals in Mayurbhanj District of Odisha' by M. Mohapatra, K.C. Satapathy and P. K. Patra revealed that most of the people in the study area work as daily labourers and they do not avail good health facilities, undergo inadequate diet, and live in non-hygienic atmosphere and non-sanitary facilities. It also highlights how all these factors aggregate to bring abnormal haematological profile among the studied communities which is a major health concern.

Third theme: 'Adult Nutrition, BMI and Puberty' included two chapters (9 and 10). Chapter 9 is contributed by Rachna Thakur and R.K. Gautam. It is a comparative study among population of two different ecological zones', authors have conducted cross-sectional study on a total of 446 women aged 18-60 years from six different endogamous groups of two ecological zones. The research findings highlights that the mean age at menarche varies from 13.66 years among Brahmins of Madhya Pradesh to 16.23 years among the Rajput of Uttarakhand. Similarly, variation in mean BMI ranging between 18.56 and 20.71 among these women. The chapter highlighted the problem of undernutrition and obesity among adult females of two different ecological zones. Further, there is information about the correlation of age at menarche and BMI.

The second chapter under the Third theme: 'Adult Nutrition, BMI and Puberty' is Chapter 10, which contributed by Harshdeep Dhanjal. The chapter describes adult nutrition among a central Indian tribe. On the basis of body mass index, chronic energy deficiency was assessed among the studied population. The chapter concludes that the females of Mawasi tribe are weaker than the males and poor nutrition shows its prevalence more among females.

The fourth theme is focused on 'Physical activity and body composition' there are two chapters (11 and 12) under this theme. Chapter 11 entiled 'Habitual physical activity, body composition and cardio-respiratory health: Study among Rajput women of District Kangra, Himachal Pradesh' is contributed by renowned Biological Anthropologist Professor Satwanti, Dr. Meenal Dhall and N.K. Mungreiphy of Department of Anthropology, University of Delhi. The chapter is focused on the relationship between body composition, cardio-respiratory health and habitual physical activity among Rajput women of District Kangra, Himachal Pradesh. The study showed that most of the subjects had normal BMI, followed by underweight but there were very few overweight subjects also. Assessment from central obesity indices also showed that most of the subjects were in normal category. Blood pressure as assessed from systolic showed that most of them were pre-hypertensive but when assessed from diastolic blood pressure most of the subjects were in normal category. Most of the subjects had normal random blood sugar level.

The second chapter under fourth theme is Chapter 12, entitled 'A study of physical activity level among the Chamba youth (17-18) years of Himachal Pradesh, The chapter is based on a cross-study, information related to different activities performed by them for two consecutive days were noted with the time duration. The result shows both the boys and girls are under the category of very high PAL (physical activity level); girls have significantly higher mean PAL than the boys, contradicting most of the studies. It can be concluded that Chamba youth are leading a very active life-style, also reflected by the low level of obesity.

Fifth theme entitled 'Nutritional status and Associated Factors' consist of three chapters (13, 14 and 15). First chapter (i.e. Chapter 13) under the theme was contributed by Dr S.K. Kolay he has attempted to assess the nutritional status of the children below six years of age belonging to Kinnaura, Bhil and Dhodia tribal population groups from Himalayan, Desert and Coastal districts of Himachal Pradesh, Rajasthan and Gujarat. Based on 12 body measurements, this study revealed that children of very poor or low economic status households have the highest rates of malnutrition. One of the major finding of the study explain that the mean and range of all the 12 measurements are higher among the Kinnauras as compared to Dhodias and Bhils.

Chapter 14 is contributed by Dr J.R. Ghosh of Shantiniketan, Kolkata. He has highlighted the prevalence of under-nutrition in terms of stunting, wasting and underweight among Santal tribal children in Birbhum district of West Bengal, India. Although the study is based on small sample size, nevertheless, it reveals poor nutritional status with a higher incidence of wasting in boys and stunting in girls, among the Santal children.

Chapter 15 entitled 'Nutritional and haematological status of migrant Pano community of slums in Bhubaneswar, Odisha' is contributed by S.K. Gouda shows how the studied community is still nutritionally vulnerable. The study is based on 24-hours dietary recall method to assess dietary intakes of children and anthropometric measurement such as height and weight recorded from children. The study findings are classified according to WHO criterion (Z-score) using nutritional indices per day per unit consumption of energy and protein values and values are compared with Recommended Dietary Allowances (RDA) of ICMR. The study result reveals that the percentage of subjects suffering from Grade I, Grade II, Grade III and Grade IV were 35 per cent, 30 per cent, 17.8 per cent and 7.8 per cent respectively. It shows that half of the population belongs to low calorie range and only 30.63 per cent households consume recommended calorie per day. It also reveals that 13 per cent have low WBC counts, 26 per cent that of low RBC, 29 per cent have low Platelet counts, 23 per cent have low haemoglobin lebel.

The sixth and last theme is on 'Food Security and Healthcare'. There is a single chapter entitled 'Role of self-help groups in providing food security and healthcare with reference to children: a study in Andhra Pradesh' contributed by M. Sreevidya. The study suggests how gender programmes helps preventing discrimination against girl child, trafficking, domestic violence helps women to increase their understanding of intra-family equity issues, decisions making levels, free mobility and necessity of building safe environment.

In this way, the book is rich in variety of aspect of human growth, development and nutrition, wide range of papers focused on different issues among variety of people from North-East to South and East to West on Indian continent. Surely it would be helpful for scholars of different field like Medical sciences, peadiatrics, human nutrition, anthropology and so on. There is no such compilation having such a wide coverage from very inception to adulthood, north to south and east to west.

List of Contributors

A.N. Sharma, Professor, Department of Anthropology, Dr. H.S. Gour University, Sagar (MP), Email: profansharma@gmail.com.

Chittaranjan Mandal, Anthropological Survey of India, 27, Jawaharlal Nehru Road, Kolkata.

D.K. Adak, Anthropological Survey of India, 27, Jawaharlal Nehru Road, Kolkata.

Gautam K Kshatriya, Professor, Department of Anthropology, University of Delhi, Delhi.

Harshdeep Singh Dhanjal, Department of Anthropology, Dr. Harisingh Gour Vishwavidyalaya (A Central University), Sagar (MP).

J. Phukan, Department of Anthropology, Cotton College, Guwahati, Assam. 3

Jyoti Jhariya, Department of Anthropology, Dr. H.S. Gour University, Sagar (MP), E- mail:drjyotijhariya@rediffmail

Jyoti Ratan Ghosh, Department of Anthropology, Visva-Bharati University, Santiniketan-731235, West Bengal, India.

Kanhu Charan Satapathy, Reader, P. G. Department of Anthropology (CAS), Utkal University, Bhubaneswar, Odisha. Email:kanhucharansatapathy @gmail.com

Lakhwinder Kaur, Department of Anthropology, Panjab University, Chandigarh

M. Sreevidya, Post Doctoral Fellow, Department of Anthropology, SV University, Tirupati. Email: sree.vinay@gmail.com

Manoranjan Mohapatra, Department of Anthropology, Utkal University, Bhubaneswar.

Meenal Dhall, Department of Anthropology, University of Delhi, Delhi.

Nitamoni Bharali, Department of Anthropology, Assam University, Diphu campus, P.O.: Diphu, Karbi Anglong, Assam 782 462, India.

Nitish Mondal, Department of Anthropology, Assam University, Diphu campus, P.O.: Diphu, Karbi Anglong, Assam 782 462, India.

NK Mungreiphy, Department of Anthropology, University of Delhi, Delhi.

P. Bharati, Biological Anthropology Unit, Indian Statistical Institute, 203, B.T. Road, Kolkata.

P. K. Patra, Department of Anthropology, Utkal University, Bhubaneswar.

P. Kashyap, Department of Anthropology, Cotton College, Guwahati, Assam.

R.K. Gautam, Department of Anthropology, Dr. Harisingh Gour Vishwavidyalaya (A Central University), Sagar (MP), Email: goutamraj@rediffmail.com.

Rachna Thakur, Department of Anthropology, Dr. Harisingh Gour Vishwavidyalaya (A Central University), Sagar (MP).

Ramesh Sahani, Department of Anthropology, Panjab University, Chandigarh

Rongkehai Ronchehon, Department of Anthropology, Assam University, Diphu campus, P.O.: Diphu, Karbi Anglong, Assam 782 462, India.

S. Jibonkumar Singh, Professor, Department of Anthropology, Manipur University, Email: drjibonkumar@yahoo.com

Satwanti Kapoor, Professor, Department of Anthropology, University of Delhi, Delhi.

Subhendu K. Acharya, Department of Anthropology, University of Delhi, Delhi.

Sunil Kumar Gouda, Department of Anthropology (CAS), Utkal University, Vani Vihar, Bhubaneswar, Odisha.

Swapan Kumar Kolay, Associate Professor & Head , School of Studies in Anthropology & Tribal Studies, Bastar University, Jagdalpur: 494005, CG.

T. Baruah, Department of Anthropology, Cotton College, Guwahati, Assam.

Thokchom Ursa, Department of Anthropology, Manipur University, Email: ursathokchom9@gmail.com

Thoudam Bedita Devi, Department of Anatomy, Sikkim Manipal Institute of Medical Sciences, Tadong, Sikkim, E mail thnoubi@gmail.com

U. Borah, Department of Anthropology, Cotton College, Guwahati, Assam.

Theme – 1

Human Growth and Development

Embryo and Foetal Growth

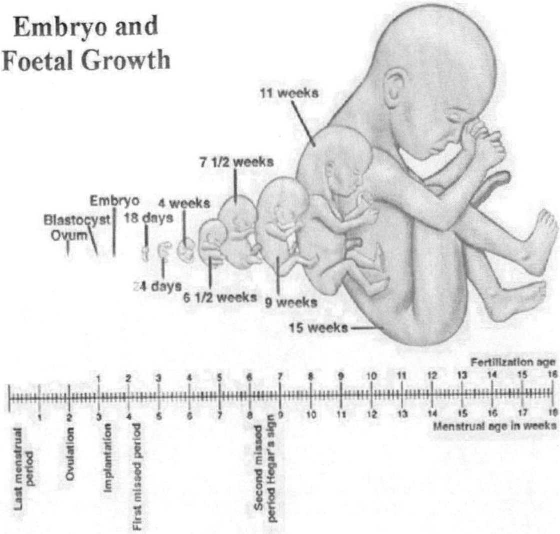

Foetal Growth from 8 to 40 Weeks

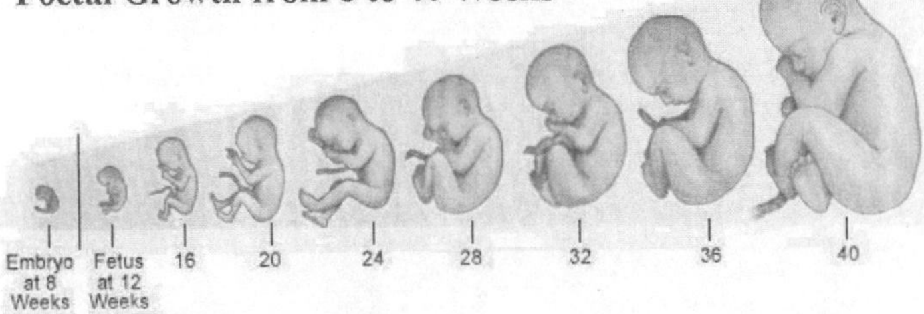

1

Ultrasonographic Prenatal Growth Study of Human Foetuses among the Meitei of Manipur

—*Thokchom Ursa, Thoudam Bedita Devi and S. Jibonkumar Singh*

Abstract

Ultrasonographic study of foetal growth is a non-invasive and effective method which can be applied for studying prenatal growth. In the present study, different stages of growth during prenatal period, starting from the embryonic period of sixth week of gestation till the fully term foetal period of 40th week are taken into account. The mean findings of the various prenatal growth parameters of the foetuses, namely; Crown –Rump length, Biparietal Diameter, Head Circumference, Abdominal Circumference and Femoral Length showed increment with increasing gestational age and the findings can represent as a reference value of foetal biometric growth chart of the Meitei foetuses. The reason for the progressive secular growth inside utero had been taken as due to genetic, change in maternal lifestyle, nutrition and better access to antenatal care programmes.

Introduction

The beginning of human life has always inspired wonder and curiosity. Birth marks not a beginning but a stage in life's journey. Quoting the famous first sentence from the chronicles of David Copperfield "To begin my life with the beginning of my life, I record that I was born" (Dickens, 1978). Indeed, it's true that human development does not start until after birth, but before taking birth. Every human being's biological beginning is a spilt-second event when a single spermatozoa which is one amongst millions of sperm cells from the father, fuses an ovum (egg cell), which is one of the several hundred thousand ova produced and stored in the mother's body. The phenomenon of fusion of the sperm with the egg cell is called fertilisation and leads to the creation of a zygote which consists of a single cell. The zygote duplicates itself again and again by the process of cell division developing into an embryo and then to a foetus by the eight week of conception and finally emerges as a complex human being consisting trillions of cells. The whole journey and process of the growth and development of the foetus during the prenatal period (the period from fertilisation until birth) takes place when the male and female sex gametes fuse together inside the female reproductive universe (oviduct or fallopian tube) where a series of remarkable events takes place making the conceptus into a full grown human being. Using LMP (Last Menstrual Period) method, a full term human pregnancy is considered to be 40 weeks (280 days) though pregnancy lengths between 38-42 weeks are considered normal (Tonse *et al.*, 2004). In the past, gestational age was established by a combination of the historical information and the physical examination. However, both the history and the findings on physical examination are fraught with error, even in the best of circumstances (Campbell *et al.*, 1985). Physical examination and historical method tends to be inaccurate and erroneous (Beazley and Underhill,1970; Hertz *et al.*, 1978; Rahilly and Muller, 1997). Mongelli et al., (1996) had ascertained that ultrasound method of estimating gestational age is better than menstrual history in predicting the date of spontaneous delivery. Wilcox *et al.*, (2000) also had same view that ultrasound dating were useful for those women with uncertain LMP and is also valid for pregnancies with reliable information of a regular LMP.

Monitoring of foetal growth is a standard component of antenatal care. The introduction of ultrasonography to obstetrics by Ian Donald and colleagues in 1958 is regarded as one of the major milestone of modern medicine (Johnson, 1998). For the first time, it becomes possible to obtain information about the foetus in its environment directly with a non-invasive diagnostic procedure which is considered safe even when used repeatedly. With the advancement in the field of scientific technology ultrasonography has become a basic diagnostic tool in obstetrics and is a part of standard component of antenatal care. Its benefits extend from use in diagnosis of very early pregnancy, assessment of a series of measurements of the foetus at different stages of pregnancy, foetal growth restriction and anomalies to estimation of foetal weight at the time of delivery. In the last 40 years, more than fifty million women have had diagnostic obstetrics ultrasound in utero and several epidemiologic studies have reported no increase in the incidence of foetal death, abnormality, intrauterine growth restriction or childhood malignancy with as much as 12 years follow up (Kremkav, 1983). Ultrasonographic study of foetal growth is a non-invasive and effective method which can be applied for studying prenatal growth. It has been 40 years since ultrasound was first used on pregnant women and unlike X-rays, ionizing radiation is not present and embryo toxic effects associated with such irradiation are not relevant. The standard ultrasound procedures provide diagnostic findings that make possible the management of problems arising during pregnancy. Upto fifty per cent of mothers, who claim to know their obstetric dates with certainty commit an error of more than two weeks when gestational age is calculated with ultrasound. A discrepancy of two weeks can be critical for the survival of a newborn that has to be delivered early because of some antenatal complications (Palmer, 1995). Conventional and Doppler ultrasound are used to assess foetal wellbeing through foetal biometry and biophysical profile (Harman, 2003). In 2005, the World Health Organisation concluded that impaired foetal growth had many causes related to: genetic factors, maternal characteristics such as nutrition, lifestyle including smoking, age, stress and disease; complications of pregnancy; and the physical, social and economic environment. More importantly proper foetal growth is a marked indication of proper maternal health and of proper balance diet during pregnancy. More specifically, inclusion of nutrient rich food, basically required by pregnant women is necessary so that the foetal growth is maintained progressively during the entire gestational period. All the reflection of the maternal lifestyle and environmental factors can be observed in the foetal growth which is diagnosed by the ultrasonographic method. Ultrasound examination of various foetal parameters also gives the information about the gestational age, biometry and well being about the foetus. In short, assessment of gestational age is fundamental to obstetric care and should be a carefully thought-out process. Assessment should depend on history and physical examination, as well as ultrasound evaluation.

The advent of ultrasound has allowed a more direct means of assessing foetal structures and development. Measurements of a wide variety of parameters have been devised to establish gestational age. Ultrasound assessment of gestational age is feasible in a majority of pregnancies and may be used to establish gestational age with greater accuracy than physical examination. The morphology is studied and various measurements are being taken which are being compared with standard age elaborated tables of several key measurements with their range listed for each gestational week. After the comparison, the ultrasound software gives the prenatal age of the embryo or foetus. Although morphological features are taken into consideration, in the first trimester, crown-rump length measurements have become the primary means of evaluating gestational age (Drumm *et al.*, 1976). In the second and third trimesters, fetal head, body, and extremity measurements have been commonly used to assess gestational age. Those parameters most commonly measured include biparietal diameter, (Campbell,1969) head circumference, (Doubilet and Greenes,1984) abdominal circumference, (Hadlock *et al.*, 1982) and femoral length (Jeanty *et al.*,1984). Therefore, considering the importance of ultrasonographic methods in studying foetal growth pattern, the present study is carried out with an objective of thorough understanding of the prenatal growth pattern of Meitei foetuses as no such study had ever been carried out and also to set up a reference value of foetal biometric growth chart of the Meitei foetuses. Meities are a dominant caste group settling in Manipur state of north-eastern part of India. They are also found settling in different states of India and even in Myanmar.

Materials and methods: The study was carried out on 460 pregnant women (singleton uncomplicated pregnancies) who attended for antenatal obstetrics ultrasound check up at the Regional Institute of Medical Science, Imphal and Langol View Maternal Clinic, Imphal. The age of pregnant women ranges from 18 to 43 years. Pregnant women with condition like hypertension, diabetes, cardiovascular condition, infections and other pregnancy-related complicacies were excluded from being the subject of study.

The following five foetal measurements were taken using a real time ultrasound scanner with a 3.5 MHz curvilinear transducer (probe), following the guidelines of the International Society of Ultrasound in Obstetrics and Gynecology (ISUOG), 2010.

1. Crown- Rump Length (CRL) 6-13th wks,
2. Biparietal diameter (BPD) 14-40th wks,
3. Head circumference (HC) 14-40th wks,
4. Abdominal Circumference (AC) 14-40th wks,
5. Femoral Length (FL) 14-40th wks.

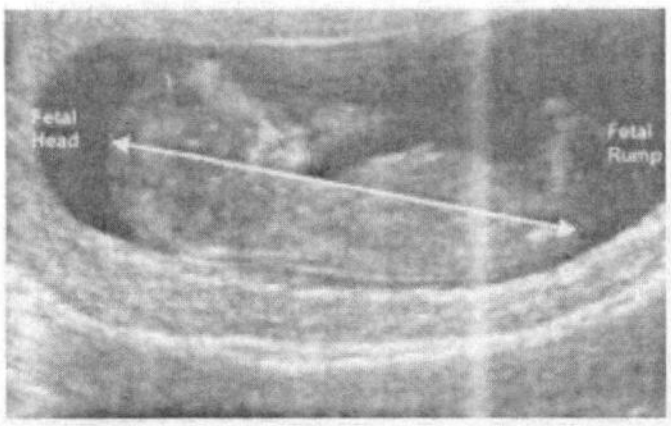

Figure 1.1: Crown- Rump Length (CRL) at 6-13th weeks.

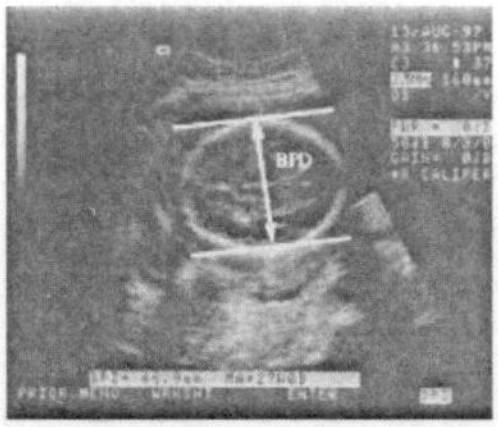

Figure 1:2. Biparietal diameter (BPD) at 14-40th weeks.

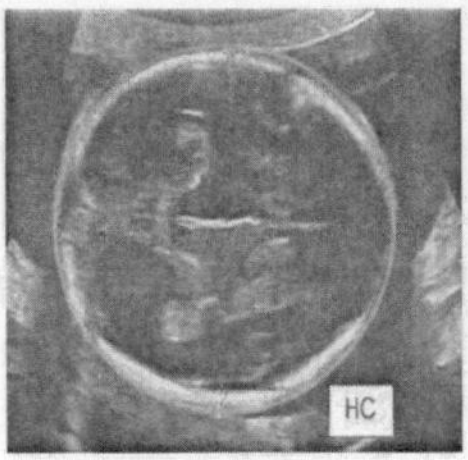

Figure 1.3: Head circumference (HC) at 14-40th weeks.

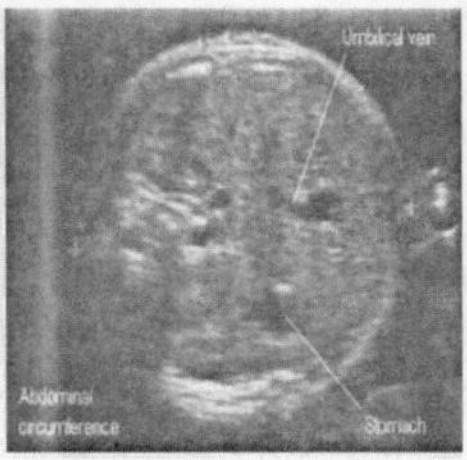

Figure 1.4: Abdominal Circumference (AC) 14-40th wks.

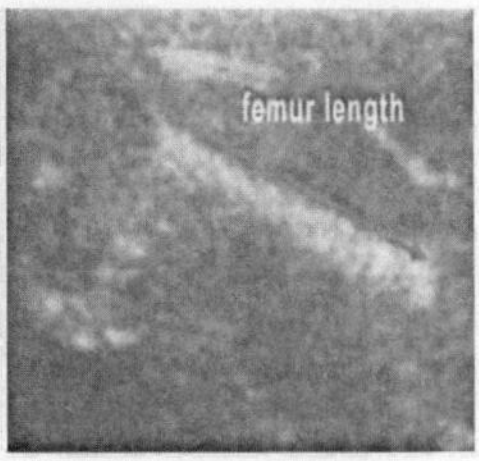

Figure 1.5: Femoral Length (FL) at 14-40th weeks.

Procedure of Measurements from ISUOG, 2010

For measuring the Crown-Rump Length, the length of the embryo's body is measured precisely from the crown of the head to the outer edge of the rump. For measuring the Biparietal diameter, the cross-sectional view of the foetal head at the level of the thalami is taken and when the continuous midline echo from the falx cerebri is broken in middle by the cavum septi pellucid and thalamus, the calliper is placed from the outer table of the proximal skull to the inner table of the inner distal skull. The Head Circumference is measured by using the same landmark as that of the Biparietal diameter, except that the calliper should be of ellipse measuring capacity. The Head circumference is then measured directly by placing the ellipse around the outside of the skull bone echoes. For measuring the abdominal circumference, the transverse section of the foetal abdomen as circular as possible is taken with the umbilical vein at the level of the portal sinus and stomach bubble visualised, the ellipse calliper is put at the outer surface of the skin line of the transverse section. The femoral length is measured by imagining optimally both ends of the ossified metaphysic clearly and then the calliper is placed at the ends of the ossified diaphysis without including the distal femoral epiphysis.

Results and Discussion

The fundamental idea behind ultrasonography of fetus at early stage is foetal growth assessment and to assess the gestational age correctly from the beginning. For this purpose, growth of the foetuses is determined through assessment of a series of the important measurements at different periods of pregnancy. During the early period of pregnancy, measurement of the Crown-Rump length between 7 to 13 weeks gives very accurate estimation of the gestational age and is helpful in dating the maturity of an embryo or early foetus. But from the 14th week onwards, the Crown-Rump length is replaced by the measurements of other body parts, namely the Biparietal diameter, the Head circumference, abdominal circumference and femoral length as they are more reliable in assessing the gestational age and the correct trend of foetal growth. The findings presented in Table 1.1 reveals a positive increasing trend of mean growth rate of the Crown-Rump length with corresponding gestational age among the studied Meitei foetuses. The Crown-Rump length measuring 6.2 mm by 6th week.

Table 1.1: Normal Range of Mean and Percentile Distribution of Crown Rump-Length (mm)

Gestational Age (in weeks)	*Mean*	*S.D.*	*Percentiles*		
			5th	*50th*	*95th*
6	6.2	2.4	3.0	6.35	11.0
7	12.5	4.9	5.1	12.0	19.9
8	19.1	3.0	14.5	19.8	22.4
9	27.3	5.0	21.6	26.0	29.7
10	36.2	4.8	30.9	37.1	44.4
11	44.5	3.7	40.9	43.0	49.8
12	61.7	6.8	54.8	60.6	68.6
13	75.6	6.0	64.8	76.8	84.0

Table 1.2: Normal Range of Mean and Percentile Distribution of Biparietal Diameter (mm)

Gestational Age (in weeks)	*Mean*	*S.D.*	*Percentiles*		
			5th	*50th*	*95th*
14	27.0	3.0	23.1	27.0	31.3
15	29.6	1.6	28.1	29.6	31.4
16	34.7	2.1	33.0	33.9	38.4
17	37.8	2.0	35.3	37.1	40.8
18	41.5	2.2	39.2	41.6	43.8
19	45.0	1.7	41.7	45.3	46.9
20	47.1	1.6	44.7	46.9	49.4
21	51.3	3.2	48.5	50.8	55.1
22	57.1	3.9	53.2	55.7	64.2
23	57.8	4.0	55.1	56.4	64.3
24	60.2	1.1	58.8	59.9	61.9
25	62.2	0.8	61.0	62.2	65.6
26	65.9	1.2	64.3	65.6	67.7
27	69.4	1.2	67.6	69.8	70.8
28	71.3	0.5	70.8	71.3	72.0
29	75.2	1.8	72.3	75.2	77.7
30	77.0	2.5	73.2	76.9	81.0
31	78.9	1.6	77.0	78.6	81.3
32	81.9	2.5	78.8	81.7	85.7

(Contd...)

33	83.6	2.2	81.7	83.0	87.0
34	85.5	2.4	82.1	85.6	88.7
35	87.6	1.5	85.4	87.9	89.4
36	89.9	1.2	88.3	89.7	91.5
37	92.0	0.7	90.6	92.0	93.4
38	93.8	1.1	91.4	94.0	95.2
39	95.0	1.3	93.3	95.4	96.4
40	95.6	3.4	91.4	97.0	97.9

Measurements of the foetal biparietal diameter and head circumference are used in assessing foetal growth, dating pregnancies and in the detection of foetal abnormalities (Chitty *et al.*, 1994). Table 1.2 and 1.3 shows the mean and percentile distribution of the Biparietal diameter and Head circumference among the Meitei Foetuses. In both the tables, a positive increase of growth rate in both Biparietal diameter and Head circumference with increasing gestational age can be seen till the 40th week. The mean value of 95.6 mm of the Biparietal diameter and 348.9 mm of Head circumference at the 40th week of gestation shows maturity of the foetuses in their growth.

Table 1.3: Normal Range of Mean and Percentile Distribution of Head Circumference (mm)

Gestational Age (in weeks)	*Mean*	*S.D.*	*Percentiles*		
			5th	*50th*	*95th*
14	102.7	11.4	89.7	102.6	119.8
15	108.0	4.7	104.0	106.9	113.7
16	127.2	6.6	118.2	126.0	134.3
17	138.8	5.5	130.0	140.0	146.3
18	157.1	7.8	149.5	156.6	165.5
19	166.4	4.9	157.9	168.2	170.5
20	177.6	6.7	169.5	178.3	187.4
21	189.0	2.5	186.1	187.9	192.9
22	204.8	4.7	198.2	205.1	211.2
23	211.6	5.0	204.3	211.8	217.3
24	224.5	3.4	218.7	225.9	228.5
25	231.0	2.9	227.3	232.3	234.8

(Contd...)

26	244.4	5.7	235.0	245.9	250.8
27	255.7	5.7	249.4	256.4	262.7
28	262.7	3.1	259.0	262.9	266.1
29	273.3	6.7	262.0	272.9	282.2
30	282.6	8.8	274.3	282.4	295.0
31	286.9	3.0	281.9	287.2	290.1
32	293.4	4.8	284.9	293.2	299.3
33	305.0	6.8	297.8	304.2	314.7
34	309.9	7.3	303.1	310.0	318.3
35	316.2	5.5	306.2	317.6	323.9
36	322.6	4.3	316.7	322.4	329.6
37	329.8	3.1	324.8	330	333.7
38	334.7	3.3	329.8	335.1	339.0
39	342.8	4.3	338.1	342.3	348.7
40	348.9	2.2	346.3	349.5	350.8

The study of biparietal diameter of foetus in obstetrics and anti-natal care with the help of ultrasound in utero has afforded modern obstetricians to know beforehand the relationship between the size and shape of the foetal head and pelvic brim. Biparietal diameter is the most important measurement of feotal head as its values are significantly superior to that of the average cranial circumference (Hadlock *et al.*, 1982). In a recent study (Mador, *et al.*, (2011) had stated that the study of biparietal diameter nearly forecast whether spontaneous delivery through the vagina would be possible or not thereby helping obstetricians to take right decision during labour and delivery.

Table 1.4 shows the mean values and per centile distribution of the Abdominal circumference of the Meitei foetuses. The Abdominal measurement reflects not only the gestational age but more of the foetal size and weight. Chamberlain (1991) had studied that serial measurement of the abdominal circumference give good warning of placental insufficiency. In the present study, the mean values of the Abdominal circumference from the 14th to 40th week increase in an affirmative manner with the increasing gestational age reflecting increment in size and weight of the foetus during the prenatal growth period which in turn can reflect better maternal nutrition and antenatal care.

The average abdominal circumference at 40th week of gestational age is 367.6 mm.

Table 1.4: Normal Range of Mean and Percentile Distribution of Abdominal Circumference (mm)

Gestational Age (in weeks)	*Mean*	*S.D.*	*Percentiles* 5th	50th	95th
14	84.7	10.3	74.3	81.1	101.1
15	94.5	4.7	89.8	93.7	106.5
16	104.7	7.3	98.4	102.5	116.2
17	118.1	7.7	108.4	117.0	130.2
18	131.3	5.2	125.9	131.3	136.8
19	144.3	6.4	136.7	143.9	153.8
20	153.2	4.1	147.8	153.8	158.3
21	164.5	4.6	158.2	164.2	172.2
22	174.0	5.8	165.7	174.3	180.8
23	184.6	6.0	176.2	184.2	192.1
24	197.3	4.8	189.6	196.9	202.9
25	206.1	8.2	197.9	205.0	219.2
26	216.9	7.3	203.8	219.0	225.0
27	229.5	4.8	224.2	229.7	236.2
28	237.0	6.3	228.3	239.1	241.4
29	251.4	4.4	246.4	251.2	257.7
30	256.6	12.2	228.6	260.1	268.5
31	272.3	5.6	266.1	274.0	279.8
32	281.7	4.4	276.1	282.4	287.5
33	293.0	8.2	285.3	290.7	303.8
34	302.0	23.9	278.7	300.0	323.8
35	313.7	7.6	300.2	313.3	325.5
36	323.6	6.2	311.3	324.2	331.2
37	336.0	4.9	330.7	334.6	343.2
38	344.7	5.9	335.1	344.1	353.0
39	350.3	7.4	341.3	351.4	358.4
40	367.6	13.6	351.4	371.5	378.3

Table 1.5 reveals the mean values and percentile distribution of the femoral length of the Meitei foetuses. The study of the measurement of femoral length during prenatal period gives knowledge about the

longitudinal growth of the foetus and allows a check on the somatic growth of the foetus (Chamberlain, 1991). In the table, the femoral length corresponds positively with the increasing gestational age and therefore can be evidence of proper growth and development during the prenatal period. The average femoral length range between 15.1 mm at the gestational age of 14th week to 78.3 at the 40th week of gestational age.

Table 1.5: Normal Range of Mean and Percentile Distribution of Femoral Length (mm)

Gestational Age (in weeks)	*Mean*	*S.D.*	*Percentiles*		
			5th	*50th*	*95th*
14	15.1	2.0	12.7	15.0	18.8
15	17.2	1.6	15.7	17.0	19.0
16	21.9	4.9	19.4	21.2	28.2
17	24.5	1.6	22.5	24.6	27.0
18	28.1	1.4	26.7	27.9	29.7
19	30.0	0.8	28.8	30.1	31.2
20	33.2	1.3	31.6	33.2	35.4
21	36.1	1.0	34.5	36.1	37.4
22	39.4	1.1	37.5	39.6	40.7
23	41.6	1.1	40.7	41.1	44.1
24	43.4	1.6	41.1	43.7	45.6
25	46.2	1.7	43.6	46.3	48.0
26	48.4	1.5	46.3	48.1	50.3
27	51.2	1.4	49.8	50.5	53.4
28	53.5	0.7	52.6	53.4	54.3
29	55.3	2.1	51.8	55.9	57.7
30	59.7	4.8	54.2	58.5	68.4
31	60.7	3.4	58.3	59.5	66.1
32	62.5	1.6	60.2	62.3	66.5
33	64.6	1.2	63.2	64.2	67.1
34	67.1	1.5	65.8	67.0	69.6
35	69.5	1.2	68.1	69.6	71.9
36	71.1	1.1	69.2	71.1	72.8
37	72.8	1.6	70.2	72.9	74.2
38	75.2	1.1	73.6	75.1	76.7
39	76.8	1.2	75.1	76.7	78.0
40	78.3	0.9	77.1	78.6	79.0

Conclusion

Obstetrics ultrasound has become an indispensible part of antenatal care starting from the screening of foetal structural anomalies to growth monitoring and its guidance to foetal interventional procedures. In the present study, the Meitei foetuses showed a positive secular nature of prenatal growth for which the factors behind may be varied. The reason for the progressive secular growth inside utero had been taken as due to genetic, change in maternal lifestyle and nutrition and better access to antenatal care programmes. The present findings of the foetal biometric growth parameters may be used as a reference value of foetal growth chart of the population as there is a need to accumulate a basic biometric data for every biologically varying population.

References

Beazley, J.M. and Underhill R.A. 1970. Fallacy of the fundal height. *Br Med J* 4: 404.

Campbell, S. 1969. The Prediction of Foetal Maturity by Ultrasonic Measurement of the Biparietal Diameter. *J Obstet Gynaecol Br Commonw* 76: 603.

Campbell, S., Warswof, S.L., Little, D., and Cooper D.J. 1985. Routine ultrasound screening for the prediction of gestational age. *Obstet Gynecol* 65: 613.

Chamberlain, G. 1991. Checking for Foetal Wellbeing. *British Medical Journal* 302(6780): 837-839.

Chitty, L.S., Altman, D.G., Hendeson, A and Campbell, S 1994. Charts of foetal size. *British Journal of Obstretics and Gynaecology.*101 35-34.

Dickens, C. 1978. *The Personal History of David Copperfield (1850).* Works of Charles Dickens. Avenel Books. New York.

Doubilet, P.M. and Greenes, R.A. 1984. Improved Prediction of Gestational Age from Foetal.

Drumm, J.E., Clinch, J. and MacKinzie, G. 1976. The Ultrasonic Measurement of Foetal Crown-Rump Length as a Method of Assessing Gestational Age. *Br J Obstet Gynaecol* 83: 417.

Hadlock, R.P., Deter, R.L., Harrist, R.B., and Park, S.K. 1982. Foetal Biparietal Diameter: A critical re-evaluation of the relation to menstrual age by means of real time Ultrasound. *Journal of Ultrasound and Medicine* (1) 97-104.

Harman, C.R. and Baschat, A.A. 2003. Comprehensive assessment of Fetal Wellbeing: Which Doppler Tests should be Performed. *Curr Opin Obstet Gynecol.* 15: 147-157.

Head Measurement. *AJR* 142: 797.

Hertz, R.H., Sokol, R.J. and Knoke, J.D. 1978.Clinical estimation of gestational age: Rules for Avoiding Preterm Delivery. *Am J Obstet Gynecol* 131: 395.

Jeanty, P., Rodesch, F., Delbeke, D. and Dumont, J.E. 1984. Estimation of Gestational Age from Measurement of Foetal Long Bones. *J Ultrasound Med* 3: 75.

Johnson, J.M. 1998. Overview of Obstetric Sonography. 'In' Carol M. Rumack, Stephanie R.

Kremkav, W.F. 1983. Biological Effects and Possible Hazards. *Clin Obstet Gynecol.* 10: 395.

Mador, E.S., Ekwempu, C.C., Mutihir, J.T., Adoga, G.I., and Ogunranti, J.O. 2011. Ultrasonographic Biometry: Biparietal Diameter of Nigerian foetuses. *Nigerian Medical Journal.* 52(1): 41-44.

Mongelli, M. Wilcox, M. and Gardosi, J. 1996. Estimating the Date of Confinement: Ultrasonographic Biometry Verses Certain Menstrual Dates. *Am. J. Obstet Gynecol.* 174:278-281.

Palmer, P.E.S. 1995. Manual of Diagnostic Ultrasound. World Health Organisation, Geneva.

Rahilly, R.O. and Muller, F. 1997. Gestational age? *Clinical Anatomy.* 10:367.

Salomom, L.J., Alfirevic, Z., Berghella, V., Bilardo, C., Hernandez-Andrade, E., Johnsen, S.L., Kalache, K. *et al.,* 2010. Practical Guidelines for Performance of the Routine Mid-Trimester Foetal Ultrasound Scan. *Ultrasound obstetrics and Gynecology.* Wiley Online Library. DOI: 10.1002/ UOG.8831.

Tonse, N.K., Barrington, K.J., Riley, L.E. and Tomashek, K.M. 2004. Age Terminology. Pediatrics. 114(5): 1362-1364.

Wilcox, A.J., Dunson, D. and Baird, D.D. 2000. The Timing of the "Fertile Window" in the Menstrual Cycle: Day Specific Estimates from a Prospective Study. *BMJ.* 322:1259-1262.

World Health Organisation. 2005. Report on the Regional Consultation Towards the Development of a Strategy for Optimizing Foetal Growth and Development. WHO Regional Office for the Eastern Mediterranean. Cairo.

2

Growth Pattern of stature and Body Weight during Childhood and Adolescence: A Cross-sectional Study among School Boys of Guwahati, Assam, India

—*T. Baruah, P. Kashyap, J. Phukan, U. Borah, D.K. Adak and P. Bharati*

Abstract

This study makes an attempt to examine the growth pattern among school boys of Guwahati city aged 5 to 16 years. Here only height and weight of the school boys (5 to 16 years of age) were considered. Altogether 265 boys were measured. Age of the boys was determined from the school records. Subjects who looked apparently healthy and active (but not on the basis of bodily structure or proportion) were included in the study. Efforts were made to exclude those with physical deformities. The children are found to be taller and heavier than the Indian urban boys. On the other hand the Guwahati school boys are much shorter and lighter than NCHS growth reference. Majority of the boys belong to the category of healthy weight. However, the frequency of underweight boy ranges between 30.77 and 42.86 per cent among these children.

Introduction

Growth studies among the children have always occupied a very prominent position in the scientific research curriculum and have always received serious attention of the researchers engaged in the field of both medical science and physical anthropology all over the world (Sharma, 1970). The growth of children in a population reflects their nutritional status and indirectly determines their living standard in general. Growth is also influenced by dietary pattern, expenditure behind food and general health condition of an individual. Slowing or cessation of growth is one of the first observable responses to nutritional inadequacy. A well designed growth study may provide a powerful tool to identify the health and nutritional status of any population or community (Tiwari *et al.*, 2007).

Many studies demonstrated that growth pattern may vary between rural and urban residence (Eveleth and Tanner, 1990 and Graham *et al.*, 1980). In India rate of growth is found to be higher in urban than in rural children (ICMR, 1989). Adair *et al.*, (1993) is of the opinion that environmental and cultural factors are responsible for urban-rural difference. In view of the above present study is an attempt to examine the growth pattern on Guwahati school boys.

Materials and Methods

Material for the present study was collected from different educational institutions of Guwahati town. This study is part of a major research project. Here only height and weight of the school boys (5 to 16 years of age) were considered. Altogether 265 boys were measured. Age of the boys was determined from the school records. Subjects who looked apparently healthy and active (but not on the basis of bodily structure or proportion) were included in the study. Efforts were made to exclude those with physical deformities.

Standing height was measured to the nearest cm using anthropometric rod (manufactured by Harpenden). Weight was measured with a physician's beam balance scale to the nearest 0.5 kg. To assess child nutrition percentile values of BMI for age were calculated. In this connection CDC growth chart developed by the National Centre for Health Statistics, USA (2000) was consulted.

Results and discussion

Cross-sectional growth of height of Guwahati school boys is presented in Table 2.1 and Figure 2.1 respectively. The mean values of the measurements increased steadily from 5 to 16 years. Highest increment in stature is found between 6-7 years (8.85 cm) and next highest is in between 12-13 years (6.6 cm) and then in 13-14 years (6.33 cm). Total increment between 5 and 16 years is 56.44 cm. It is observed that there are five growth spurts between the ages of 6 and 7 years, 8 and 9 years, 12 and 13 years, 13 and 14 years and 14 and 15 years in Guwahati school boys. However, the highest is observed between the ages 6 and 7 years. Per cent growth per annum is found to be highest in 7 years (8.11 cm) and lowest is in 16 years (1.25 cm).

Table 2.1: Age-wise Stature (cm) among school boys of Guwahati (Assam), India

Age groups (in years)	*No.*	*Mean±S.E.*	*S.D. ±S.E.*	*Difference between mean (cm)*	*Per cent growth per annum*
5	13	106.07±0.31	1.13±0.22	-	-
6	13	109.15±0.47	1.69±0.33	3.08	2.90
7	20	118.0±1.32	5.92±0.94	8.85	8.11
8	20	122.54±0.86	3.84±0.61	4.54	3.85
9	29	128.64±0.86	4.62±0.61	6.10	4.98
10	45	133.37±1.06	7.09±0.75	4.73	3.68
11	38	137.09±0.82	5.06±0.58	3.72	2.79
12	29	141.32±1.56	8.40±1.10	4.23	3.08
13	14	147.92±1.69	6.32±1.19	6.60	4.67
14	19	154.25±0.78	3.39±0.55	6.33	4.28
15	12	160.50±0.50	1.72±0.35	6.25	4.05
16	13	162.51±0.54	1.95±0.38	2.01	1.25

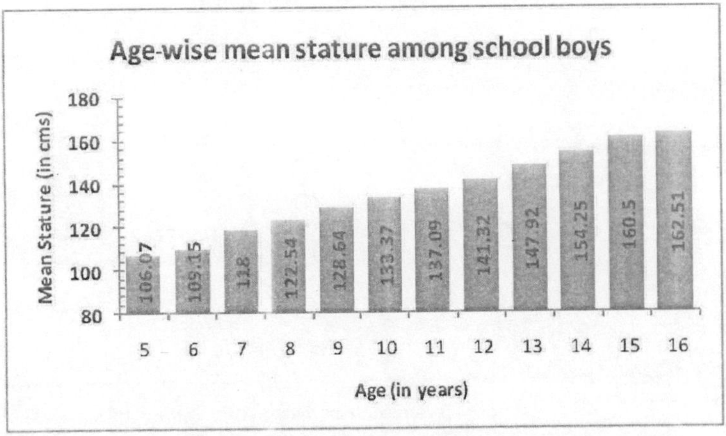

Figure 2.1: Age-wise mean stature among school boys of Guwahati (Assam) India.

Growth of weight of Guwahati school boys is presented in Table 2.2 and Figure 2.2 respectively. Like stature mean values of weight increased steadily from 5 to 16 years. Highest increment in weight is found between 13-14 years (7.89 kg) and next highest is in between 6-7 years (4.8 kg) and then in 15-16 years (4.44 kg). Total increment between 5 and 16 years is found to be 31.84 kg. Altogether

Table 2.2: Age-wise weight (kg) among school boys of Guwahati (Assam), India

Age groups (in years)	*No.*	*Mean±S.E.*	*S.D. ±S.E.*	*Difference between mean (kg)*	*Per cent growth per annum*
5	13	15.85±0.65	2.34±0.46	-	-
6	13	16.85±0.55	1.99±0.39	1.00	6.31
7	20	21.65±0.80	3.60±0.57	4.80	28.49
8	20	22.70±0.75	3.34±0.53	1.05	4.85
9	29	24.72±0.86	4.63±0.61	2.02	8.90
10	45	28.87±0.88	5.93±0.62	4.15	16.79
11	38	30.45±0.80	4.91±0.56	1.58	5.47
12	29	31.90±1.16	6.25±0.82	1.45	4.76
13	14	34.28±0.93	3.49±0.66	2.38	7.46
14	19	42.17±0.81	3.52±0.57	7.89	23.02
15	12	43.25±0.79	2.73±0.56	1.08	2.56
16	13	47.69±0.41	1.49±0.29	4.44	10.26

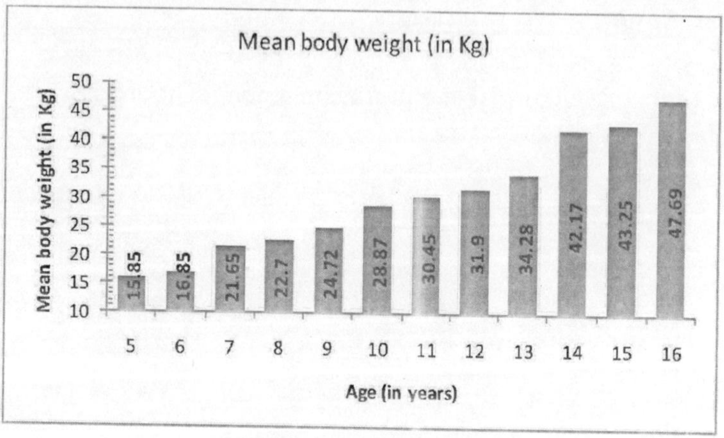

Figure 2.2: Age-wise mean body weight among school boys of Guwahati (Assam) India.

four growth spurts are observed. These are between the ages of 6 and 7 years, 9 and 10 years, 13 and 14 years and 15 and 16 years. However, the highest is observed between the ages 13 and 14 years. Per cent growth per annum is found to be highest in 7 years (28.49 kg) and lowest is in 15 years (2.56 kg).

Comparison of stature and weight of the present study with that of Indian urban boys

Stature: It is seen from Table 2.3 that Guwahati school boys are taller than the Indian urban boys at each group from 5 to 16 years. The difference is found to be small in 6, 13 and 16 years. The mean value of stature, is however, little higher among Indian urban boys (109.3 cm) than that of the Guwahati school boys (109.15 cm).

Table 2.3: Comparison of stature (cm) of present study with Indian Urban boys (ICMR, 1989)

Age in years	*Guwahati school boys*	*Indian urban boys*
5	106.07	102.7
6	109.15	109.3
7	118.0	114.5
8	122.54	119.8
9	128.64	124.1
10	133.37	128.7
11	137.09	133.9
12	141.32	138.8
13	147.92	145.3
14	154.25	151.4
15	160.50	156.8
16	162.51	160.3

Weight: It may be seen from Table 2.4 that Guwahati school boys are heavier than the Indian urban boys at each group from 5 to 16 years. The difference is noted to be small in 5 and 6 years. The mean value of weight, is however, much higher among Guwahati school boys than that of the Indian urban boys in 10 and 14 years of age.

Table 2.4: Comparison of weight (kg) of present study with Indian Urban boys (ICMR, 1989)

Age in years	*Guwahati school boys*	*Indian urban boys*
5	15.85	15.0
6	16.85	16.6
7	21.65	18.4
8	22.70	20.0
9	24.72	21.8
10	28.87	23.8
11	30.45	26.4
12	31.90	29.0
13	34.28	32.7
14	42.17	36.8
15	43.25	40.9
16	47.69	44.0

Comparison of stature and weight of the present study with that of NCHS

Stature: Unlike Indian urban boys it is seen that Guwahati school boys are much shorter than NCHS growth reference (Table 2.5) at each group from 5 to 16 years. The difference is comparatively lower in 5, 7 and 8 years than other ages. While the total increment between 5 and 16 years is 56.44 cm in case of present study it is 61.9 cm in case of NCHS.

Table 2.5: Comparison of stature (cm) of present study with NCHS (1979)

Age in	*Present Study*			*NCHS*		
years	*N*	*Mean*	*SD*	*N*	*Mean*	*SD*
5	13	106.07	1.13	675	112.6	3.0
6	13	109.15	1.69	298	119.2	3.7
7	20	118.0	5.92	312	125.1	4.2
8	20	122.54	3.84	296	129.8	5.2
9	29	128.64	4.62	322	135.8	6.3
10	45	133.37	7.09	334	140.9	7.8
11	38	137.09	5.06	324	146.4	10.0
12	29	141.32	8.40	349	152.2	11.1

(Contd...)

13	14	147.92	6.32	350	159.2	11.6
14	19	154.25	3.39	359	167.1	11.9
15	12	160.50	1.72	359	170.8	11.2
16	13	162.51	1.95	349	174.5	11.9

Weight: Like stature means of weight of Guwahati school boys are much lower than NCHS growth reference (Table 6) at each group from 5 to 16 years. However, the difference is comparatively lower in 5, 6, 7 and 8 years than other ages. While the total increment of weight between 5 and 16 years is 31.84 kg in case of present study it is 41.1 kg in case of NCHS.

Table 2.6: Comparison of weight (kg) of present study with NCHS (1979)

Age in years	*Present Study* *N*	*Mean*	*SD*	*NCHS* *N*	*Mean*	*SD*
5	13	15.85	2.34	676	19.9	5.3
6	13	16.85	1.99	298	22.6	5.4
7	20	21.65	3.60	312	25.1	5.7
8	20	22.70	3.34	296	27.7	6.3
9	29	24.72	4.63	322	31.3	5.8
10	45	28.87	5.93	334	35.4	6.9
11	38	30.45	4.91	324	39.8	7.4
12	29	31.90	6.25	349	44.2	8.1
13	14	34.28	3.49	348	49.6	8.8
14	19	42.17	3.52	359	56.9	8.2
15	12	43.25	2.73	359	61.0	7.3
16	13	47.69	1.49	349	66.8	7.1

Child nutrition

To assess child nutrition (5-16 years), height and weight for 265 boys are measured. Percentile values of BMI for age were calculated. In this connection CDC growth chart developed by the National Centre for Health Statistics, USA was consulted. BMI according to percentile values were furnished in Table 2.7.

Table 2.7: BMI for age (6-14 years)

Age in years	PERCENTILE VALUE			
	<5th (Underweight)	*5th-85th (Healthy weight)*	*85th-95th (Over weight)*	*≥ 95th (Obese)*
5	4 (30.77)	8 (61.54)	1 (7.69)	-
6	4 (30.77)	9 (69.23)	-	-
7	1 (5.0)	16 (80.0)	2 (10.0)	1 (5.0)
8	3 (15.0)	16 (80.0)	1 (5.0)	-
9	11 (37.93)	15 (51.72)	2 (6.90)	1 (3.45)
10	10 (22.22)	29 (64.44)	4 (8.89)	2 (4.44)
11	9 (23.68)	22 (57.89)	5 (13.16)	2 (5.26)
12	10 (34.48)	17 (58.62)	2 (6.90)	-
13	6 (42.86)	8 (57.14)	-	-
14	3 (15.79)	16 (84.21)	-	-
15	1 (8.33)	11 (91.67)	-	-
16	1 (7.69)	12 (92.31)	-	-

Figures in parenthesis indicate percentage values.

Frequency of underweight boy is considerably high in 5, 6, 9, 12 and 13 years, which ranges between 30.77 and 42.86 per cent. Majority of the boys belong to the category of healthy weight. Per cent of overweight boys is considerably low. In 7 (10 per cent) and 11 years (13.16 per cent) considerable per cent of the boys found to be overweight. It is interesting to note that in 6, 13, 14, 15 and 16 years no overweight boy is found. However, very few of the boys are found to be obese. They are represented in 7, 9, 10 and 11 years only. Altogether six boys are found to be obese. Thus, in fine it can be said that though the Guwahati school boys are heavier and taller than our national average they are lighter and shorter than NCHS reference.

References

Adair, L. S., J. Vanderslice and N. Johoori. 1993. Urban-rural difference in growth and diarrhoeal morbidity of Filipino infants. In: L.M. Schell, M. Smith and A. Bilsborough (Eds.): Urban ecology and health in the Third World. Cambridge University Press, Cambridge.

Eveleth, Ph. B. and J. M.Tanner. 1990. World-wide variation in human growth. Cambridge University Press, Cambridge.

Graham, G. G., W. C. Maclean, C. H. Kallman, J. Rabold, D. Mellits. 1980. Urban-rural difference in the growth of Peruvian children. *Am. J. Clin. Nutr.*, 33: 338-344.

Hamill, P.V.V., T. A. Drizd, C. L. Johnson, R. B. Reed, A. F. Roche, W. M. Moore. 1979. Physical Growth: National Center for Health Statistics percentiles. *The American Journal of Clinical Nutrition*. 32: 607-629.

Indian Council of Medical Research. 1989. Growth and Physical Development of Indian Infants and children. Technical Report Series, No.18, ICMR, New Delhi.

National Center for Chronic Disease Prevention and Health Promotion, USA. 2000

Sharma, J. C. 1970. Physical growth and development of the Maharashtrians. Ethnographic and Folk Culture Society, Lucknow.

Tiwari, M. K., K. K. N. Sharma, S. Bharati, D. K. Adak, R. Ghosh and P. Bharati. 2007. Growth and Nutritional Stauts of the Bharia – A Primitive Tribe of Madhya Pradesh. *Coll Antropol*, 31 (1): 95-101.

3

Pattern of Growth among Homozygous Sicklers: Cross-sectional Study among Mehra a Scheduled Caste of Mandla District, Madhya Pradesh

—Jyoti Jhariya, Awadhesh N. Sharma and Rajesh K. Gautam

Abstract

This paper highlights the problems faced by homozygous individuals of sickle cell disease. The findings are based on empirical field investigation among Mehra a scheduled caste of Mandla District, Madhya Pradesh. A cross-sectional study was carried out on homozygous and normal individuals of age 2-20 years. A total of 99 individuals were studied out of which 7 were homozygous for the sickle cell gene. Anthropometric measurements were taken on these individuals during the summer of year 2012. They were measured for weight (W), height (H), sitting height (SH), waist circumference and hip circumference using the standard technique. It was found that the homozygous children have delayed growth as compared to normal children. Further, the homozygous children have smaller trunk as compared to normal children. Present findings indicate that the homozygous children for sickle cell gene have delayed growth and development at all ages, especially after 5 years of age. Height for age is 15 per cent delayed as compared to normal children, whereas body weight for age is 46 per cent delayed. They have stunting as well as wasting.

Introduction

Sickle cell anemia is a genetic health problem widespread among Indian population. It is one of the heamoglobinopathy. It is lethal in homozygous condition. The children with homozygous condition hardly survive to reproductive age, but due to advancement of medical health- care their life expectancy is increased and some of them may survive upto 30-40 years of age with several complications (Jhariya *et al.*, 2015). It was found that they face multiple problems, acute pain, priyaprism, alienation, constant hospitalisation, economic constraints etc. There is no exclusive policy for such deprived children.

Sickle cell disease is widely spread among the people of tropical countries and transmitted as autosomal recessive character. If a person receives only one gene responsible for sickle hemoglobin from either of the parent, the condition is called carrier or trait. If one inherits two defective genes, one from each parent, the condition is called sickle cell disease. As the carrier state provides protection against malaria, it has attained high frequency in many parts of the tropical world (Jhariay *et al.*, 2015). Person with trait leads a normal life but the diseased person suffers from various complications throughout the life such as anemia, bone and joint pain, swelling on joints, recurrent infection, osteomyelitis, necrosis of bone, aplastic crises, abdominal pain, splenic sequestration crises, hepato-splenomegaly etc (Serjeant and Serjeant, 2001). Ultimately it shortened the life expectancy.

Platt *et al.*, (1994) reported an average life expectancy of 42 in males and 48 in females. It is an autosomal co-dominant genetic blood disorder characterised by red blood cells that assume an abnormal, rigid, sickle shape. Sickling decreases the cells' flexibility and results in a risk of various complications. In India, the presence of sickle gene was first reported by Lehman and Cutbush (1952). The gene of sickle cell was occurred because of a mutation in the Hemoglobin gene. In medical terminology it is called Sickle-cell disease (SCD), or sickle-cell anemia (or anemia; SCA) or drepanocytosis.

Growth is a fundamental and common property of life. Its course is governed by certain fundamental laws. Human body growth is a complex process which has been defined by Garn (1952) as a change in magnitude, increments in the size of the organs, increases in the thickness of tissue or change in the size of an individual as a whole. Macy and Kelly (1957) defined it as the biochemical metabolic relationship between physical and chemical composition of the body. Human body growth is governed by certain factors, such as nutrition, climate, socio-economic condition, diseases and genetic structure. The most important among these is nutrition. Available data in this field suggest that sub-optimal nutrition during infancy and early childhood delays growth. Many scientists (Jackson, 1966; Behar, 1968, Tanner, 1962 and other) are of the opinion that when adequate diets of children were supplemented with nutrition the rate of growth improved. Children born to mother of sufficient nutrition, enriched food regarding proteins, vitamins during pregnancy were better in general health than the children born from mothers on poor diet (Chaurasia, 1986).

The growth is also governed by genetic constitution of an individual. The interaction of genetics and environment is well established as 'Nature-Nurture'. There are studies on growth pattern of Indian children belonging to tribal, rural and urban backgrounds (Mitra, 2002; Gautam, 2007a and 2007b; Ahirwar and Gautam, 2015; Thakur and Gautam, 2014 and 2015); but there is no information about the growth pattern of homozygous children for sickle cell gene. The reason behind is that the homozygous children were hardly survive upto 10-15 years of age. Then, most of researchers have focused in the finding out the prevalence of gene rather than growth pattern of children having sickle cell gene (Papiha *et al.*, 1987; Gautam and Sharma, 2002 and 2003; Patel and Athavale, 2004). Another important aspect of studying homozygous children was their clinical complications (Serjeant and Serjeant, 2001 and Yadav *et al.*, 2006). Therefore, this study is exclusively focused on the growth pattern of homozygous children for sickle cell gene.

Material and Method

For present investigation cross-sectional study was carried out on homozygous and normal individuals of age 2-27 years belonging to a scheduled caste known as Mehra of Mandla district of Madhya Pradesh. A total of 99 individuals were studied out of which seven were homozygous for the sickle cell gene. Detailed case studies were recorded of these homozygous children, adolescents and adults, for details Jhariya *et al.*, (2015) can be referred. To assess the pattern of growth and development quantitatively anthropometric measurements were taken on these individuals during the summer of year 2012. They were measured for weight (W), height (H), sitting height (SH), waist circumference and hip circumference using the standard technique (Martin and Saller, 1956). Accuracy of the data is properly taken care. To calculate cormic index (SH/H), BMI (W/H2) and waist-hip ratio MS-Excel and SPSS software were used. Height for age, weight and BMI is drawn using MS-Excel. Cormic index and body mass index, waist hip ratio were calculated for each individuals followed by calculation of central tendency viz. mean and deviation of each of measurement. Following formulas are used:

$$\text{BMI} = \frac{\text{Weight (in kg)}}{\text{Height (in sq meter)}}$$

$$\text{Cormic Index} = \frac{\text{Sitting Height (in cms)}}{\text{Height (in cms)}}$$

$$\text{Waist Hip Ratio} = \frac{\text{Waist Circumference (in cms)}}{\text{Hip Circumference (in cms)}}$$

Results

Age-wise height, sitting height, body weight, waist girth, hip girth, BMI, cormic index and waist hip ratio of homozygous (for sickle cell gene) children of 5-18 years were measured as shown in Table 3.1. It is evident that at five years of age the homozygous children is 110 cm tall, he has body weight 13 Kg and BMI of 10.65 Kg/m^2 with waist hip ratio 1.00 and cormic index 0.471. Gradual increment in height, body weight, waist girth, hip girth and BMI of such children can be observed from the Table. At the same time there is gradual decline in waist hip ratio as per progression of age can be witnessed.

Table 3.1: Age-wise anthropometric measurements and indices of Homozygous (HbSS) children for sickle cell among Mehra a Schedule Caste of Mandla District (MP) (After Jhariya *et al.*, 2015)

Age	*Height*	*Sitting Height*	*Weight*	*Waist girth*	*Hip girth*	*BMI*	*Cormic Index*	*Waist Hip ratio*
5	110.5	52.0	13	46	46	10.65	0.471	1.000
7	108.0	50.5	16	50	51	13.72	0.468	0.980
9	114.8	59.0	17	51	55	12.90	0.514	0.927
11	120.3	59.4	20	60	63	13.82	0.494	0.952
14	158.6	64.6	30	64	74	11.93	0.407	0.865
18	161.0	79.2	45	71	79	17.36	0.492	0.899

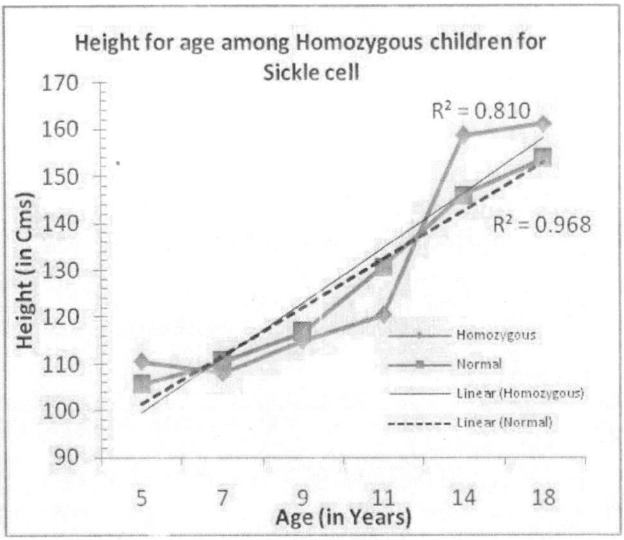

Figure 3.1: Comparative Height for age among Normal and Homozygous children(After Jhariya et al., 2015).

Cormic index is a ratio of sitting height and height in other words it is a ratio of length of trunk and total body length. Higher value of cormic index indicate larger trunk as compared to total body length or stature. In the present study it was found that the homozygous children have smaller trunk as compared to normal children except one exception at age of 9 years.

Age-wise descriptive statistics of anthropometric measurements and indices for normal individuals of Mehra caste of 2 to 20 years of

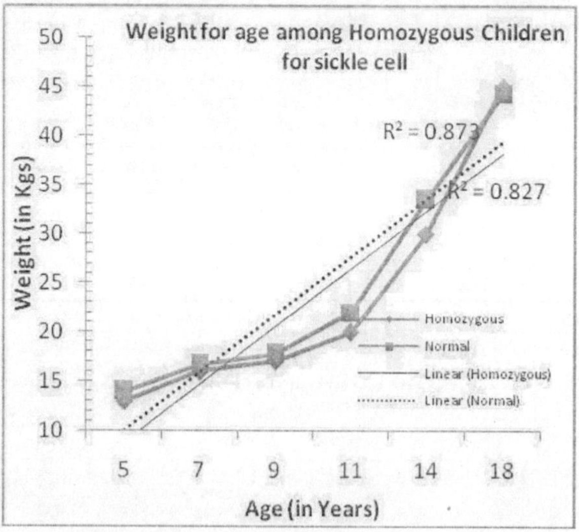

Figure 3.2: Comparative weight for age among Normal and Homozygous children (After Jhariya et al., 2015).

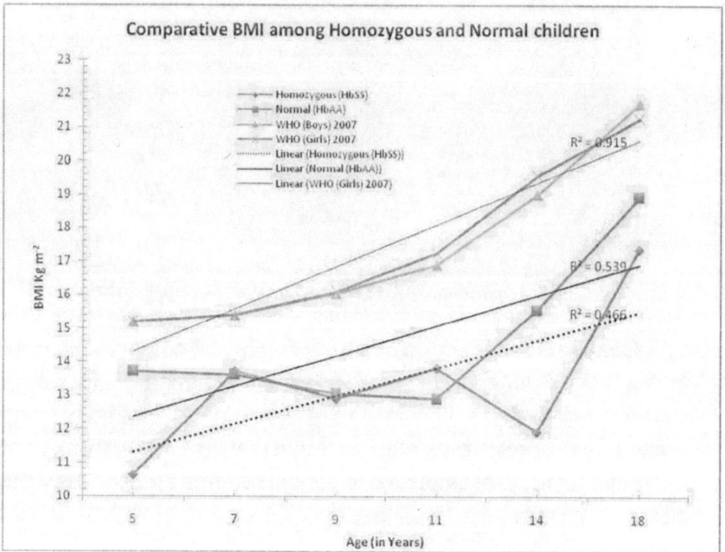

Figure 3.3: Comparative BMI among Homozygous and Normal Children. (After Jhariya et al., 2015).

Table 3.2: Age-wise mean and SD of Anthropometric measurements and Indices of Normal children among Mehra caste of Mandla district (MP) (After Jhariya *et al.*, 2015)

Age	N	Stature		Sitting height		Weight		Waist girth		Hip girth BMI		Cormic Index		Waist Hip		Ratio	
		Mean	SD	Mean	SD	Mean	SD	Mean	SD	Mean	SD	Mean	SD	Mean	SD	Mean	SD
2	2	72.8	3.9	40.5	7.8	9.5	0.7	41.5	0.7	42.0	2.8	17.96	0.59	0.555	0.077	0.99	0.06
3	3	91.7	1.0	51.1	1.9	11.0	0.0	45.0	2.0	47.3	1.2	13.10	0.30	0.557	0.015	0.95	0.05
4	4	93.5	3.9	51.1	2.5	12.3	1.3	48.5	1.3	50.5	1.0	13.99	0.71	0.405	0.270	0.96	0.02
5	3	105.5	6.0	62.7	15.1	14.0	1.0	47.0	1.4	51.5	7.8	13.71	0.86	0.666	0.202	0.84	—
6	3	108.1	4.5	58.1	2.1	11.7	2.9	49.7	4.7	54.0	4.4	9.89	1.64	0.538	0.022	0.92	0.05
7	3	110.5	2.2	55.2	4.9	16.7	0.6	63.0	21.7	50.3	5.0	13.61	0.08	0.515	0.033	0.99	0.08
8	3	119.5	5.5	60.0	1.6	18.7	3.1	49.7	2.9	56.7	2.9	13.01	1.13	0.503	0.011	0.87	0.01
9	6	116.7	4.8	61.0	4.6	17.8	2.9	50.6	3.6	63.2	17.5	13.05	1.32	0.412	0.232	0.82	0.21
10	4	135.8	23.4	64.1	1.5	21.0	2.0	53.3	2.1	65.5	13.0	11.83	2.61	0.481	0.071	0.83	0.11
11	5	130.6	14.8	65.3	3.9	22.0	2.7	56.6	3.1	61.6	1.8	12.89	2.25	0.505	0.043	0.91	0.04
12	4	135.2	6.0	67.2	3.6	26.0	4.2	58.0	2.9	68.5	5.0	14.19	1.70	0.498	0.025	0.85	0.04
13	5	139.8	12.2	71.1	5.7	32.0	10.3	54.2	17.1	70.6	9.7	16.02	2.73	0.509	0.013	0.76	0.21
14	10	146.0	8.2	72.7	5.2	33.5	7.1	63.2	6.2	74.6	7.4	15.56	2.10	0.499	0.036	0.85	0.05
15	6	151.1	13.1	77.3	6.6	39.5	10.1	57.0	19.8	75.8	5.6	17.06	1.85	0.512	0.011	0.75	0.25
16	10	153.4	5.8	77.9	4.3	41.6	4.8	66.4	6.1	80.7	3.3	17.68	1.62	0.508	0.015	0.82	0.07
17	4	158.8	10.2	79.5	3.5	45.8	6.8	69.3	6.2	82.5	7.5	18.12	1.59	0.502	0.012	0.84	0.09
18	9	153.6	10.3	76.9	5.5	44.2	7.4	68.0	6.6	81.1	3.2	18.92	2.61	0.502	0.012	0.83	0.06
19	4	155.9	12.9	77.9	5.3	42.3	6.1	65.0	8.0	81.3	3.0	17.38	1.48	0.501	0.018	0.80	0.08
20	4	158.9	7.8	79.5	4.7	52.8	12.3	74.5	12.9	87.0	5.5	20.72	3.30	0.501	0.014	0.85	0.11
Total	92																

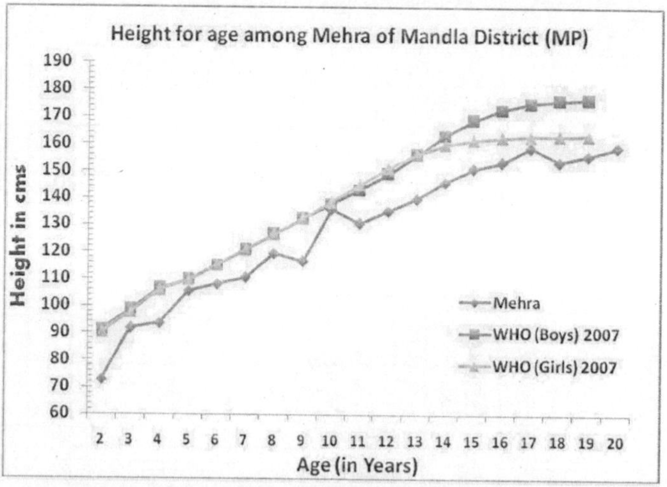

Figure 3.4: Height for age among Mehra of Mandla District (MP) (After Jhariya et al., 2015).

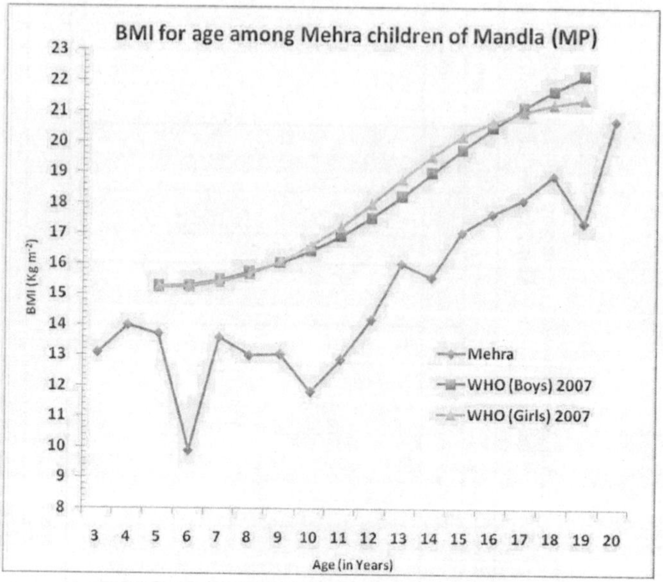

Figure 3.5: BMI for age among Mehra children of Mandla (MP) (After Jhariya et al., 2015).

age are presented in Table 3.2. It is evident that height, weight, waist girth, hip girth and BMI of children grow as per growing age. The mean stature at 2 years of age was 72.8±3.9 cms with body weight 9.5±0.7 Kg, waist girth 41±0.7 cms, hip girth 42±2.8 cms and BMI 17.96±0.59. They have cormic index 0.555 and waist hip ratio 0.99. At age of 20 years, they have attained an average height of 158.9±7.8 cms with sitting height of 79.5±4.7 cms, having body weight of 52.8±12.3 Kg, waist girth 74.5±12.9 cms, hip girth 87±5.5 cms, BMI 20.72±3.30 Kg/m^2, cormic index 0.501±0.014 and waist hip ratio 0.85±0.11.

For better understanding of differential growth pattern among Homozygous sicklers and normal individuals comparative line graph are drawn for height-for-age and weight-for-age as shown in Figure 3.1 and 3.2. It is evident that the homozygous children have delayed growth as compared to normal children. For further elucidation of facts comparative line graph are drawn for age-for-BMI as presented in Figure 3.3. Similarly, line graph is drawn for for height-for-age and weight-for-age (Figure 3.4 and 3.5) to compare normal children with WHO (2007) reference data. It is apparent that the homozygous children have delayed growth as compared to normal one; at the same time the normal children having delayed growth as compared to WHO (2007) reference data for boys and girls.

Conclusion

The present findings indicate that the homozygous children for sickle cell gene have delayed growth and development at all ages specially after five years of age. Height for age is 15 per cent delayed as compared to normal children whereas Weight for age is 46 per cent belated. They have stunting as well as wasting. To understand the overall growth pattern among the studied population a total of 92 Normal children of 2-18 years of age were measured; and it was found that in general they have delayed growth than the WHO (2007) reference data, and the homozygous children have further delayed growth than the normal children. It can be recommended that there should be National policy to take care of such children. Alike thalassemia, sickle cell disease should also be included in the right to disability bill.

References

Behar, (1968). Cf. Chaurasia (1986). *Body growth among Jain boys and girls*. Students Book Depot. Sagar, M.P.

Chaurasia, R.K. (1986). *Body growth among Jain boys and girls*. Students Book Depot. Sagar, M.P.

Garn, S.M. (1952). Physical growth and development. *Am.J.Phys. Anthrop.* 10:169-192.

Gautam, RK. 2007a. Physical Growth and Body Composition among two male populations of Central India— Baiga (Primitive Tribe) and Brahmin. In contemporary Studies in Anthropometry *(Ed.) A.N. Sharma*, Sarup and Sons, New Delhi, pp. 81-89.

Gautam, RK. 2007b. Physical Growth and Body Composition among two female populations of Central India— Baiga (Primitive Tribe) and Brahmin. In contemporary Studies in Anthropometry *(Ed.) A.N. Sharma*, Sarup and Sons, New Delhi, pp. 139-147.

Guatam, RK and Sharma, AN (2002). Occurrence of Sickle Gene and Malaria Among the Mehra (Jhariya) Caste of Mandla District, Madhya Pradesh, India. *Anthropologist*, 4 (4): 281-283.

Guatam, RK and Sharma, AN (2003). Evaluation of Awareness Regarding Sickle Cell among Sicklers and Non-Sicklers of Mehra (Jhariya) of Mandla District, Madhya Pradesh, Central India. *South Asian Anthropologist* 3.2 (2003): 107-110.

Jackson, (1966). Cf. Chaurasia (1986).

Jhariya, J., Sharma, AN and Gautam, RK (2015). Growth and development of Homozygous Sicklers: Case Study among Mehra a Scheduled Caste of Mandla District, Madhya Pradesh. *Asian Journal of Multidisciplinary Studies*, 3, 8: 50-55.

Lehman and Cutbush 1952. Sickle cell in Southern India. *British Medical Journal*, 5 (I), 404.

Macy, I G. and Kelly, H.J. (1957). *Chemical Anthropology. A New Approach to growth in Children*. University of Chicago Press, Chicago.

Martin, R, and K Saller (1956). *Lehrbuch der Anthropologie*, Vol. 3, Stuttgart, G. Fisher.

Papiha, SS, Roberts DF, Mukherjee DP, Singh SD, and Malhotra M, (1978). A genetic Survey in the Bhil Tribe of Madhya Pradesh Central India. *Ann. Phys. Anth*, 49, 2:179-185.

Patel, AB and Athavale AM (2004). Sickle cell disease in Central India. *Ind. J. Paediatr.* 71:789-793.

Platt, OS, Brambilla DJ, Rosse WF, *et al.* (1994). "Mortality in sickle cell disease. Life expectancy and risk factors for early death". *N. Engl. J. Med.* 330 (23):

1639–44. doi:10.1056/NEJM199406093302303. ISSN 0028-4793. PMID 7993409.

Sarma, PS, Viswanathan KA, Mukherjee MM (1986). Death patterns in sickle cell disease from a steel city. *JAPI,* Vol. 34. pp. 192-194.

Serjeant, G.R. and Serjeant, B.E. (2001). *Sickle Cell Disease.* Oxford University Press, New York.

Tanner, J.M. (1962). *Growth at adolescence.* Blackwell Scientific Publications, Oxford.

Yadav, R, Gupta RB, Bharadwaj VK, Singh MPSS (2006). Morbidity Profile of Sickle Cell Disease in Central India. *In* Tribal Health proceeding of National Symposium October 19-20. Edited by Singh *et al.*, 2006. Regional Medical Research Centre for Tribals (Indian Council of Medical Research), Jabalpur, Madhya Pradesh, India.

4

Growth and Nutritional Status of Adolescent Tibetans in Odisha

—*K. C. Satapathy*

Abstract

Study on the growth and nutritional status of adolescent Tibetans born and brought up in coastal Odisha was compared with high altitude studies by using standard anthropometric technique. This study is based on cross-sectional data collected between February 2003 and September 2004. Anthropometric data on children and adolescents were collected from Central schools for Tibetans, individual's home or at community centre. Only those children, adolescents, and young adults who were born and raised in the settlement or near by town or city were included in the present study. A total of 98 males and 131 females were recruited. Height was measured with the help of Harpenden anthropometer rod and weight with a weighing scale. Height was recorded to the nearest millimeter and weight to nearest 0.5 kg. BMI cut off was used for classifying the population for nutritional status. Mercury sphygmomanometer was used to measure blood pressure. All the measurement was taken by single investigator. Age of the children and adolescents were collected from school records and were cross-checked with the individuals. The secondary data on birth weight was mostly collected from the local hospitals. In the present study it has been observed that Tibetans in costal Odisha shows a trend towards become taller and heavier when compared in terms of their height, weight, BMI, and birth weight. Blood pressure level of adolescent Tibetans shows a trend towards higher prevalence of hypertension and is similar to that of high altitude studies. In terms of Nutritional status the present generation exhibits a better picture when compared to other populations.

Key words: *Tibetans, Odisha, Height, Weight and BMI (Body Mass Index), birth weight, blood pressure and nutrition.*

Introduction

People permanently living at >3000 meter above sea level are exposed to hypoxia and show a phenotypic form of adaptation (Meere *et al.*, 1996). Growth at high altitude is generally characterised by smaller body size than low altitude (Beall *et al.*, 1977, Greksa *et al.*, 1984, 1985; Hoff, 1974; Stinson, 1980; Weitz and Garruto, 2004). Children living at high altitude often have delayed growth but whether growth retardation is related to high altitude or there are other factors behind it is still not known (Tripathy *et al.*, 2007; Argnani *et al.*, 2008). In fact, the effects of hypoxia on growth are combined with the effects of low socioeconomic status, poor nutrition and disease. The few studies conducted in Tibet indicated that there is a high prevalence of malnutrition in preschool children (especially stunting). These studies also implied that there is some adaptation to high-altitude hypoxia in the growth pattern of Tibetan children (Harris *et al.*, 2001; Dang etal, 2004; Tripathy *et al.*, 2007; Argnani; 2008; Rooze *et al.*, 2012). Considerable variation in newborn size exists within and between populations. At high altitude, women of high altitude ancestry give birth to heavier babies compared to high altitude residents who were born and raised at low altitude or with women born and raised at high altitude but descended from populations of low-altitude ancestry (Zamudio *et al.*, 1993, Niermeyer *et al.*, 1995, Hass *et al.*, 1980). Altitude also affects blood pressure. Both systolic blood pressure and diastolic blood pressure are found to lower in the high altitude native population than low altitude native populations at high altitude (Clegge *et al.*, 1976, Hanna 1999). The cause of this decline is not clear, it has been suggested that hypoxia leads to relaxation of vascular smooth muscle or that an increase in collateral circulation may be involved (Frisancho, 1993). The degree of decline in systemic Blood Pressure is found to be a function of time at residence at altitude (Hanna, 1999). The Andean evidence is strong in suggesting the above relationship.

This observation is also observed in some other populations like Sherpas (Basu *et al.*, 1984), natives of Tien Shan and the Pamir (Mirrakhimov, 1978) and the Ambaras region in Ethopia (Beall *et al.*, 1997). One study on Tibetans (Sun 1986) showing higher incidence of hypertension than the low altitude native Han migrants.

Tibetans born and raised at low altitude do not experience the multitude of stresses that exist at high altitude, hypoxia being the most important. Thus, the study of growth and nutrition of Tibetans in the absence of hypoxia at low altitude will provide an insight into the unique and possibly adaptive pattern of growth and development of Tibetans at high altitude. Tibetans are believed to have at high altitudes for longer than other high-altitude populations of the world (Moore, 1998). Dalai Lama of Tibet was given asylum in India after the political upheavals in 1959. Following Dalai Lama, 8000 Tibetans fled to India. Tibetan refugees have been living in different parts of India for more than four decades. The different parts of India offers them different climatic and cultural environment and where the succeeding generations have grown up without any stress of hypoxia. In the present study an attempt has been made to examine the degree to which their phenotypes differ from those of their parental generations at high altitude through the determinant of health i.e. height, weight, BMI, birth weight, nutrition and blood pressure. It also provides a measure of the environmental impact on the growth process to hypoxia and there by mechanism of adaptation to high altitude. For the above purpose Tibetans who are an Asian high altitude population that has migrated and settled in Odisha in an almost alien and completely different ecosystem were chosen and compared with the Tibetans of other studies in different situations.

Materials and Methods: The Tibetans settlement in Odisha was established in 1963 at Chandragiri on Eastern Ghats, which is at an altitude of 930 meter and is also very near to coastal area as well as away from the urban centres drastically different from their homeland Tibet. The settlement started with an initial population of 2500 and reached approximately to 3,600 in 2001. This study is based on cross-sectional data collected between February 2003 and September 2004. Anthropometric data on children and adolescents were collected from Central schools for Tibetans, home and community centre.

Only those children, adolescents, and young adults who were born and grown up in the settlement or nearby town or city were included in the present study. A total of 98 males and 131 females were recruited. Age of the children and adolescents were collected from school records and were cross-checked with the individuals. The secondary data on birth weight was mostly collected from the local hospitals. Anthropometric measurements were taken as described by Weiner and Lourie (1981). Height was measured with the help of Harpenden anthropometer rod and weight with a weighing scale. Height was recorded to the nearest millimeter and weight to nearest 0.5 kg. CDC/NCHS (2005) recommended BMI cut off was used for classifying the population for nutritional status. Mercury sphygmomanometer was used to measure blood pressure. All the measurement was taken by single investigator. For classification of Hypertension recommendation of JNC7 (2003) was used. For the nutritional assessment the information on dietary intake and food habits were collected, using 24 hours recall methods (Gopalan, 2000).

Results

The finding of present study is presented in Table and illustrated by graphs. The mean height of Tibetan males at different altitude is shown in Table 4.1, whereas for females is shown in Table 4.2.

Table- 1 Mean heights (cm) of Tibetans (Male) at different Altitudes

Age Group in years	*N=98*	*Chandragiri 930 m*	*Weitz 3,200 m*	*Weitz 3,800 m*
6-7	19	115.9	111.4	111.3
8-9	9	123.4	120.8	121.1
10-11	12	132.5	128.4	132.9
12-13	9	146.5	139.5	144.1
14-15	17	154.2	149.5	153.7
16-17	13	164.7	164.7	165.1
18-19	15	168.9	168	165.4
20-29	12	170.0	168.7	167.5

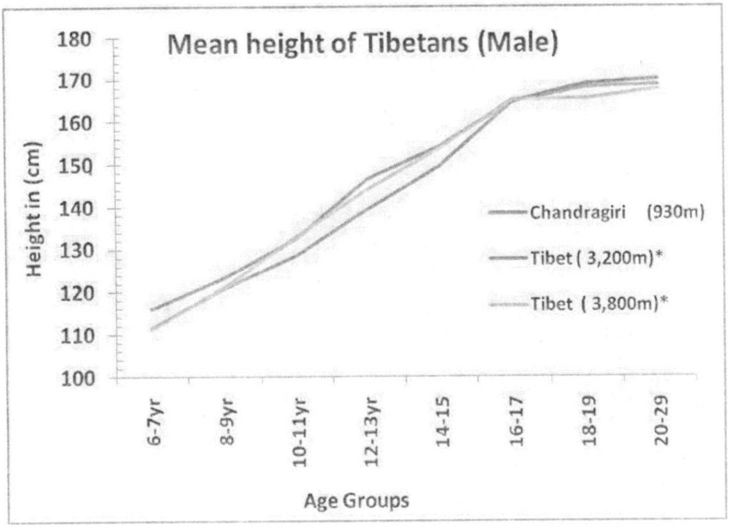

Figure 4.1: Mean heights of Tibetans (Male) at different Altitudes.

In the above Table 4.2 and Fig. 4.2 the mean heights of Tibetan females in Odisha were compared with the high altitude studies. It has been observed that mean heights of Tibetans increases in all the age groups i.e. from 6-7 yr to 20-29 yr. Except for the age group 8-9 yr and 20-29 yr the height for age value is seen to be higher in the present study as compared to high altitude studies.

Table 4.2: Mean heights of Tibetans (Female) at different altitude

Age Group in years	*N=131*	*Chandragiri 930 m*	*Weitz 3,200 m*	*Weitz 3,800 m*
6-7	11	116.6	111.9	113.5
8-9	14	124.3	120.5	125
10-11	13	135.6	128.4	137.4
12-13	21	147.9	141.5	146.9
14-15	21	153.9	150.2	151.2
16-17	19	157.9	153.8	155.3
18-19	7	157.3	156.6	155
20-29	25	155.6	157.1	157.4

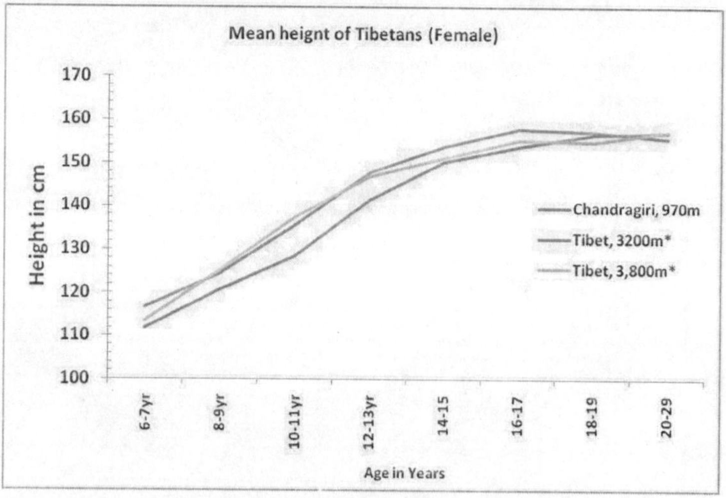

Figure 4.2: Mean heights of Tibetans (Female).

In Table 4.3 and Fig. 4.3 the mean weight (kg) of Tibetan males in Odisha was compared to the different high altitude studies and it was found that the mean value of the present study is higher than the high altitude studies for all age groups.

Table 4.3: Mean weights of Tibetans (Male)

Age Group in years	*N=98*	*Chandragiri 930 m*	*Weitz 3,200 m*	*Weitz 3,800 m*
6-7	19	21.3	18.6	18.8
8-9	9	24.4	22.4	22.3
10-11	12	27.9	26.3	27.8
12-13	9	38.0	31.1	34.3
14-15	17	42.3	38.1	43.3
16-17	13	53.7	52.3	53.3
18-19	15	58.0	56.6	55.7
20-29	12	62.9	57.5	57.8

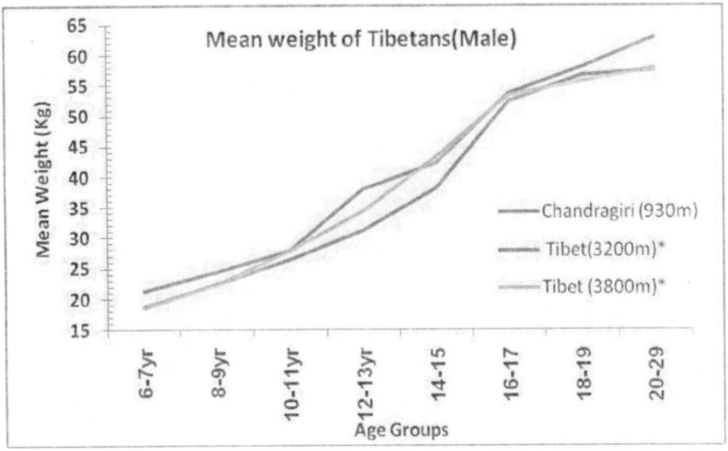

Figure 4.3: Mean weights of Tibetans (Male).

The mean weight (kg) of Tibetan females in the present study (Table 4.4 and Figure 4.4) was compared with the different high altitude studies and it was found that the mean value of the present study is higher than the high altitude studies for all age groups.

Table 4.4: Mean weights of Tibetans (Female)

Age Group in years	*N=131*	*Chandragiri 930 m*	*Weitz 3,200 m*	*Weitz 3,800 m*
6-7	11	20.8	17.9	19.2
8-9	14	24.9	20.9	22.4
10-11	13	30.5	24.8	28.7
12-13	21	43.2	31.7	36.4
14-15	21	47.3	40.5	42.6
16-17	19	52.5	47.2	50.6
18-19	7	55.9	52.1	52
20-29	25	54.4	51	50.4

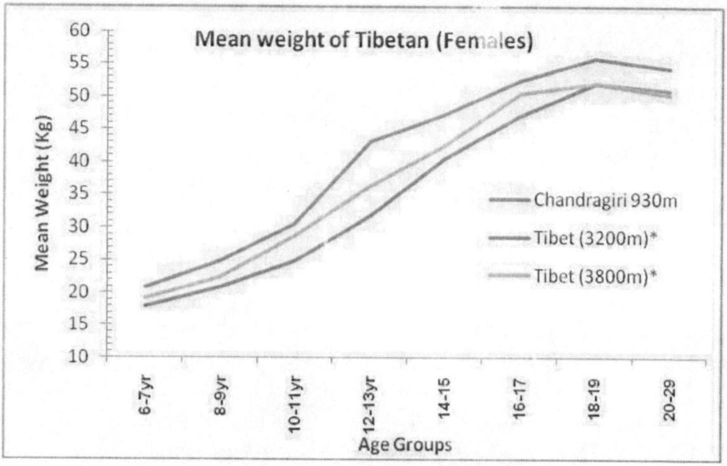

Figure 4.4: Mean weights of Tibetans (Female)

Table 4.5 present the nutritional status of Tibetan children based on height for age Z scores which shows that out of 208 individual 15.9 per cent of them are fall under stunted category whereas 84.1 per cent belongs to normal category. It has also been observed that the variation in sex is non-significant (χ^2 =0.242, P>0.05). However, high incidence of stunting has been reported among Tibetan children in high altitude.

Table 4.5: Nutritional Status based on Height for Age Tibetan (Children)

Sex	*Stunted*		*Normal*		*Total*
	N	*per cent*	*N*	*per cent*	*N*
Male	17	17.2	82	82.8	99
Female	16	14.7	93	85.3	109
Total	**33**	**15.9**	**175**	**84.1**	**208**

χ^2 =0.242, P>0.05

Table 4.6 presents the nutritional status of Tibetan children weight for age of 191 individuals shows that three per cent belongs to wasted category whereas 97 per cent belongs to normal category. It has also been found that the variation between sex is non-significant (χ^2 =0.322, P>0.05).

Table 4.6: Nutritional Status based on Weight for Age Tibetan (Children)

Sex	*Wasted*		*Normal*		*Total*
	N	*per cent*	*N*	*per cent*	*N*
Male	2	1.0	86	43.7	88(3.0)
Female	4	2.0	105	53.3	109(97.0)
Total	**6**	**3.0**	**191**	**97.0**	**197(100.0)**

$\chi^2 = 0.322$, $P>0.05$

Table 4.7 presents the nutritional status of Tibetan children born and brought up in Odisha according to weight for stature Z score value and it reveals that all i.e. 108 individuals are belongs to normal category.

Table 4.7: Nutritional Status based on Weight for Stature Tibetan (stature between 81 to 149 cm)

Sex	*Wasted*		*Normal*		*Total*
	N	*per cent*	*N*	*per cent*	*N*
Male	0	0	51	47.2	51
Female	0	0	57	57.8	57
Total	**0**	**0**	**108**	**100.0**	**108**

Body mass index is a measure of weight for height and is used differently with Children than it is with Adults. In children and in teens, body mass index is used to assess underweight, overweight, and risk for overweight. Children's body fatness changes over the years as they grow. Also, girls and boys differ in their body fatness as they mature. This is why BMI for children, also referred to as BMI-for-age, is gender and age specific. BMI-for-age is plotted on gender specific growth charts. These charts are used for children and teens 2 – 20 years of age. CDC (2005) percentile cutoff points were used in the present study to identify underweight and overweight in children as well as for adults over 20 years old.

Table 4.8 presents the BMI percentage of 218 Tibetan children, 79.4 per cent of them belong to normal category, 8.3 per cent belong to underweight category, 7.3 per cent belong to risk of overweight category and only 5 per cent of them belong to overweight. Similarly, among 98 male children 12.24 per cent belong to underweight category, 74.48 per cent belong to normal category, 6.12 belong to risk of over weight and 7.14 per cent belong to the overweight category. Among the females out of the total 120 individuals 83.33 per cent belong to normal category, 5 per cent belong to underweight, 8.33 per cent belong to the risk of underweight category and 3.33 per cent belong to the overweight category. From this it has been observed that the number of underweight cases is more in case of male as compared to the females.

Table 4.8: BMI of Tibetan Children (2-20 yrs) in Odisha

Category	*Cutoff percentile*	*Male (per cent)*	*Female (per cent)*	*Total (per cent)*
Under weight	BMI-for-age <5th percentile	12(12.24)	6(5.0)	18(8.3)
Normal	BMI-for-age>5th percentile to <85th percentile	73(74.48)	100(83.33)	173(79.4)
At risk of overweight	BMI-for-age 85th percentile to <95th percentile	6(6.12)	10(8.33)	16(7.3)
Overweight	BMI-for-age > 95th percentile	7(7.14)	4(3.33)	11(5.0)
Total		**98(45.0)**	**120(55.0)**	**218(100.00)**

In Table 4.9 the BMI percentage of 186 Tibetan adults shows that 58.06 per cent belong to normal category, 4.84 belong to underweight category 23.12 per cent belong to the risk of over being weight category and 13.98 per cent belong to obese category. Out of 69 males 57.97 per cent belong to the normal category, 7.25 per cent belong to under-weight category, 23.18 per cent are in the risk of overweight category and 11.6 per cent belongs to overweight category. Similarly, out of 117 females 58.06 per cent belong to normal category, 4.84 per cent belong to underweight category, 23.12 per cent belongs to at risk of overweight category and 13.98 per cent belongs to obese category. From this it has been observed that the number of obese cases is higher among the females as compared to males but the cases of underweight is more in case of males as compared to females. However, there is a similarity seen in both the sexes whether normal or whether at the risk of being overweight. Thus, the adult population of the Tibetans leaving in Odisha varies between normal to obese category.

Table 4.9: BMI of Tibetan Adults in Odisha

Weight Status	*BMI*	*Male (per cent)*	*Female (per cent)*	*Total (per cent)*
Underweight	>18.5	5(7.25)	4(3.4)	9(4.84)
Normal	18.5-24.9	40(57.97)	68(58.11)	108(58.06)
At risk of overweight	25.0-29.9	16(23.18)	27(23.08)	43(23.12)
Obese	30.0 and above	8(11.6)	18(15.38)	26(13.98)
Total		**69**	**117**	**186**

Table 4.10: Comparison of Mean BMI for different countries

Place	*Year*	*No. of Adults*	*BMI Mean*	*References*
India* (Males)	1988/90	9447	18.9	Shetty and James, 1994
India* (Female)	1988/90	11914	19.0	Shetty and James, 1994
Odisha, Tibetan (Male)	2003-2005	69	24.2	Tripathy & Satapathy, 2006
Odisha, Tibetan (Female)	2003-2005	117	25.2	Tripathy & Satapathy, 2006
Tibet, Tibetan Nomads (Male)	1996	19	17.7	Beall *et al.*, 1996
Tibet, Tibetan Nomads (Female)	1996	19	17.1	Beall *et al.*, 1996

The comparison of BMI of Tibetans in the present study (Table 4.10) with other populations including Tibetans in the high altitude shows that Tibetans in Odisha shows a better mean value for BMI i.e. 24.2 for males and 25.2 for females as compared to the high altitude Tibetans i.e. 17.7 in case of males and 17.1 in case of females and Indians with 18.9 in case of males and 19.0 in case of females. Therefore, the BMI value which is considered to be an indicator of nutritional status shows better value as compared to other population in the present study. This is due to change in environment and change in food and dietary habits of the present generation Tibetans in Odisha.

The above Table 4.11 shows the comparative study of nutrient intake of Tibetans at different altitude compared with that of ICMR recommended value. It has been observed that the mean calorie consumption of Tibetans adults in Odisha is 3256 Kcal as compared to high altitude study i.e. 3017 and 2343 Kcal but ICMR recommended value is 2800 Kcal. Similarly, the mean carbohydrate consumption is 557 gm in Odisha. Protein consumption of Tibetans in Odisha is 86.2 gm as compared to high altitude study i.e. 85.4 gm and 52.6 gm but ICMR recommended value is 55.0 gm. As far as fat consumption is considered Tibetans in Odisha consume 54.2 gm daily as compare

Table 4.11: Average daily nutrient intake (g) of Tibetans at different altitude

Nutrients	*Presnet Study, Tibetans (970 m), Odisha, 2006 (N=223)*	*Tibetans, Odisha (970 m), Patel, 1985*	*Tibetan Women Dehradun (600 m), Bera; 2004*	*Sherpas, Kalimpong, (1000-1500 m) Gupta et al., 1989*	*Sherpas, Upper Khumbu (3,500-4,500), Gupta et al., 1989*	*RDA, Male ICMR, 2004*
Calorie (Kcal)	3256	2629	2504	3,017	2,343	2,800
Carbohydrate (g)	557	498	424.0	-	-	330
Protein (g)	86.2	36.7	75.7	85.4	52.6	55.0
Fat (g)	54.2	6.1	51.7	52.3	-	40

to high altitude i.e. 52.3 gm and recommended value is 55.0 gm per day. It shows that the nutrient intake of Tibetans in Odisha is better then the high altitude study and is better as compared to ICMR recommended value.

Table 4.12: Results from birth weight studies in Tibetans at different altitudes

Place	*Altitude (m)*	*N*	*Birth weight (g)*	*SD*	*References*
Lhasa, Tibet	3658	34	3222	400.0	Zamudio et. al., 1993
Kathamadu, Nepal	1200	45	3313	322.6	Zamudio et. al., 1993
Tibet	>4000	96	2860		Moore *et al.*, 2001
Tibet	3000-4000	165	3140		Moore *et al.*, 2001
Tibet	>3000	116	3110		Moore *et al.*, 2001
Chandragiri, Odisha (Tibetans)	930	54	3404	499.3	Present study
Chandragiri, Odisha (Odias)	930	94	2518	534.9	Present study

Table 4.12 presents the birth weight of Tibetans at different altitude and its shows an increasing in birth weight in low altitude Odisha when compared it with the high altitude studies.

Table 4.13: Hypertension among the adult Tibetans in Odisha

Category	*SBP**	*DBP**	*Male (per cent)*	*Female (per cent)*	*Total (per cent)*
Normal	<120	And <80	26(29.54)	45(31.46)	71(30.73)
Prehypertension	120-139	Or 80-89	31(35.22)	41(28.67)	72(31.16)
Stage 1Hypertension	140-159	Or 90-99	19(21.59)	20(13.98)	39(16.88)
Stage 2Hypertension	≥ 160	Or ≥ 100	12(13.63)	27(18.88)	39(16.88)
Total			**88 (100.00)**	**143 (100.00)**	**231 (100.00)**

Table 4.13 presents the classification of hypertension according to Sex of adult Tibetans of Odisha. The percentage of normal blood pressure is 29.5 for males and 31.6 for females. Stage I and Stage II hypertension are 21.59, 13.63 and 13.98, 18.88 for males and females respectively. Percentage of adults classified as pre-hypertension is 35.22 for males and 28.67 for females which suggests an increasing trend of high blood pressure among the Tibetans in Odisha.

Table 4.14: Average blood pressure of Tibetans at different altitude

Studies	*Place and population*	*Systolic, mmHg*		*Diastolic, mmHg*	
		Male	*Female*	*Male*	*Female*
Present Study, 2004	Odisha (Tibetans)(N=328)	122.9	121.9	75.6	78.2
Sun, 1986	Lhasa (Tibetans)(N=15545)	123.7	129.5	79.7	82.5
Gupta, 1984	Kalimpong (Sherpas)	128.6	132.6	81.4	85.3
Nirmala, 2001	Indian (Rural)	104	98	74	71
	Indian (Urban)	106	90	75	65

Table 4.14 presents Tibetans blood pressure in low and high altitude. It has been observed that Tibetans have a high blood pressure values in both the altitude when compared to the local studies. However, residents at high altitude are known to have a reducing effect on systemic blood pressures.

Discussion: Growth and development is the result of environmental effect on the manifestation of genetic potential of the individual. The present status of growth and development of an individual is the result of the environmental stresses it encountered throughout its developmental period. The study of growth and development has been described in two sections, one describing prenatal growth and the other postnatal growth. Prenatal growth has been studied through birth weight which is an easily measured integrated result of prenatal growth. At high altitude, decreased growth in utero is sustained after birth. High altitude growth is characterised by slow and delayed rate of growth, less pronounced growth spurts and shorter stature compared to low altitude (Haas, 1976; Haas *et al.*, 1977; Leonard *et al.*, 1995; Beall, *et al.*, 1977; Beall 1982; Frisancho and Baker, 1970; Frisancho, 1969; Mueller *et al.*, 1978; Malik, 1987; Malik and Pandey, 1993; Malik and Singh, 1978, 1984; Ji and Ohsawa, 1993; Beall, 1984; Pawson, 1977; Frisancho 1978). In the present study, we compare the growth of Tibetans at high altitude to that of the effect of the absence of hypoxic stress by studying the growth of Tibetans at low altitude. Table 4.1 to Table 4.4 compare the mean height and mean weight respectively of Tibetans at high altitude and Tibetans at low altitude. For the comparison of our data we selected the data of Weitz (2000) from Tibet. His data were from three different places at different altitudes all above 3000 m. We pooled his data for the three places for each age group and plotted it alongside the height and weight curves of the present study. The low altitude mean height and weight of the present study were greater than the Weitz's sample.

Anthropometric indices are used as the main criteria for assessing the adequacy of diet and growth in individuals and/or for a population. BMI a composite index of weight and height is used with age and sex independent cut-offs to define overweight or thinness in adults (CDC 2005). NCHS reference percentiles cut-offs 5th, 85th and 95th as provided by Frisancho (1990) has been used for assessing nutritional status in children and adolescents (Table 4.4, 4.8, and 4.10). Present study reveals that the low altitude Tibetans have high BMI value as compared to high altitude Tibetans as well as other Indian population. The nutritional status of the Tibetan in terms of Z score as a whole show a very pleasant picture as far as under nutrition is concerned (Table 4.5 to 4.7). However, an alarmingly high percentage of adult population in the category of risk of overweight and obesity is a matter of concern.

At both low and high altitudes, Tibetans have higher birth weights compared to other populations at the same locale (Table 4.12). The difference in birth weight between low and high altitude is inversely proportional to the protection from IUGR (Intra Uterine Growth Retardation) at high altitude. Our study shows that Tibetan birth weight was more than that of groups with low-altitude ancestry who were living at the same altitudes. A new finding in this study is that Tibetan birth weights are higher among Tibetans living at low altitudes than the Tibetans living at high altitudes. The increase in birth weight with decreasing altitude among Tibetans is similar to what was observed for Andeans in Peru by Beall (1981b) and in Bolivia by Haas *et al.*, (1980).

The Tibetans' heavier birth weights at high altitude compared to persons living at the same altitudes but of low-altitude ancestry may indicate better adaptation compared to the people of low-altitude ancestry at high altitude. Otherwise, it may be due to genetic factors influencing birth weight irrespective of altitude (Tripathy, 2005). The Tibetans' heavier birth weights at low altitudes indicate that foetal growth does not achieve its potential even among Tibetans at high altitude. Genetic potential for birth weight is seemingly manifested only at low altitude. Migration of Tibetans to low altitude has resulted in an increase in blood pressure (Table 4.13 and 4.14) a possible reason could be that Tibetans at high altitude in contrast to Tibetans at low altitude have retained the habit of high salt intake and higher intake of non-vegetarian in their diet (Table 4.11).

References

Argnani, L; Cogo A.; and Gualdi-Russo E. Growth and Nutritional Status of Tibetan Children at High altitude. *Collegium Anthropologicum*. 2008, 32: 807-812.

Baker, P. T. Human adaptation to high altitude. *Science*. 1969, 163: 1149-1156.

Basu, A and Gupta R. 1989. Ed. Human biology of Asian highland populations in the global context. *Indian Anthropological Society*, Calcutta.

Beall, CM. 1976. The effect of high altitude on growth, morbidity and mortality of the Peruvian infants. PhD thesis, Pennsylvania State University, Pennsylvania, USA.

Beall, CM. 2000. Tibetan and Andean patterns of adaptation to high altitude hypoxia. *Human Biology* 72(1): 201-228.

Beall, CM. 2004. Higher offspring survival among Tibetan women with high oxygen saturation genotypes residing at 4,000 m. *PANAS* 1073(10):1-5.

Beall, CM. 1981a. Growth in a population of Tibetan origin at high altitude. *Annals of Human Biology* 8(1): 31-38.

Beall, CM and Goldstein MC. 1981b. Tibetan fraternal polyandry: a test of sociobiological theory. *American Anthropologist* 72(1): 201-228.

Beall, CM and Goldstein MC. 1987. Hemoglobin concentration of pastoral nomads permanently resident at 4,850-5,450 meters in Tibet. *American Journal of Physical Anthropology* 73:433-438.

Bellezza, J. V. 2001. Antiquities of Northern Tibet: Pre-Bugdhist archaeological discoveries on the High Plateau. Adroit Publishers, Delhi.

CDC/NCHS. 2005. Center for disease control and prevention. Department of Health and Human Services. USA. Website: http://www.cdc.gov/nccdphp/dnpa/bmiadult.htm (Retrieved June13, 2005).

Clegg, EJ, Harrison GA, Baker PT. 1970. The impact of high altitudes on human populations. *Human Biology* 42:486-18.

Dang, S, Yan H, Yamamota S, Wang X and Zeng L. 2004. Poor nutritional status of younger Tibetan children living at high altitude. *European Journal of Clinical Nutition*. 58: 938-946.

Frisancho, AR. 1978. Human growth and development among high altitude populations. In: Baker PT. 1978. Ed. The Biology of high-altitude people. IBP Synthesis, Volume 14. Cambridge University Press. Cambridge.

ICMR. 2004. Nutrient requirements and recommended dietary allowances for Indians. A report of the expert group of the Indian Council of Medical

Research. New Delhi: Indian Council of Medical Research (ICMR), pp. 1-334.

Ghai, O.P. 1980. Maternal factors in the epidemiology of low birth weight. *Indian Journal of Pediatrics*. 47:123-128.

Gopalan, C, Ramasastri BV, Balasubramanian SC. 2000. Nutritive values of Indian foods. National Institute of Nutrition, Hyderabad, (ICMR).

Gupta, R. 1980. Altitude and demography among the Sherpas. *Journal of Biosocial Science*. 12: 103-14.

Gupta, R. 1981. The impact of altitude on human population. PhD thesis, Calcutta University, Calcutta.

Gupta, R, Basu A, Pawson IG, Bharati P, Mukhopadhyay S, Roy SK, Majumdar PP, Bhattacharya SK, Bhattacharya KK, Das SK. 1989. Altitude and Human Biology: A Comparative Study of Himalayan, Andean and Ethiopian Data.In: Human Biology of Asian Highland Population in the Global Context. Ed Basu A and Gupta R. Pp. 1-80. 9 *Indian Anthropological Society* Occasional Papers, Calcutta.

Gupta, SP and Ramachandran KS. 1995. History of Tibet. Tibetan Parliamentary and Policy Research Center, New Delhi.

Hann, JM. 1999. Climate, altitude, and blood pressure. *Human Biology*. 71:553-82.

Hass, JD, Frongillo EA, Stepick CD, Beard JL, Hurtado L. 1980. Altitude, ethnic and sex difference in birth weight and length in Bolvia. *Human Biology*, 52:459-77.

ICMR. 1972. Growth and physical Development of Indian Infant and Children. ICMR Technical Report. Ser. No. 18.

JNC 7. 2003. The seventh report of the Joint National Committee on Prevention, Detection, Evaluation, and Treatment of High Blood Pressure. US Department of Health and Human Services. NIH Publication No. 03- 5233.

Lahiri, S, Edelman NH, Cherniak NS, Fishman AP. 1969. Blunted hypoxic drive to ventilation in subjects with life-long hypoxemia. Federation Proceedings 28:1289.

Malik, SL. 1987. Physical growth and altitude: comparison of high and lowlander Bods of the Western Himalayas. *Indian Anthropologist*.17:9-48.

Malik, SL. 1976. Age changes and few biological responses at high altitude, a study among the male Bods of Ladhak (PhD Thesis, unpublished). Delhi University, Delhi.

Mazess, RB. 1975. Biological Adaptation: Aptitude and Acclimatisation. In: Biosocial Interaction in population Adaptation (Ed) Watt EC *et al.*, pp. 9-18. Mountain Publishers.

Moore, LG, Niemeyer S, Zamudio S. 1998. Human adaptation to high altitude: regional and life cycle perspectives. *American Journal of Physical Anthropology Suppl* 41: 25-64.

Moore, LG, Young D, Mccullough RE, Droma T, Zamudio S. 2001. Tibetan protection from ntrauterine growth restriction (IUGR) and reproduction loss at high altitude. *American Journal of Human Biology* 13:635-644.

Niermeyer, S, Yang P, Shanmina, Drolkar, Zhuang J, Moore LG. 1995. Arterial oxygen saturation in Tibetan and Han infants born in Lhasa, Tibet. *N. Engl. J. Med.* 333:1248-52.

Nirmala, A. 2001. Age variation in blood pressure: effect of sex and urbanisation in a genetically homogeneous caste population of Andhra Pradesh. *American Journal of Human Biology*. 13(6): 744-52.

Oni, GA. 1986. The effect of maternal age, education and parity on birth weight in a Nigerian Community: The comparison of results from bivariate and multivariate analysis. *Journal of Tropical Pediatrics* 32:295-300. (Cited by Mondal B, 1997).

Pawson, IG. 1974. The growth and development of high altitude children with special emphasis on populations of Tibetan origin in Nepal. Ph.D Thesis. Pennsylvania State University. USA.

Relethford, JH. 1990. Studies of human biological variation and adaptation. *Review in Anthropology* 15: 167-70.

Rooze, S, Dramaxi-Wilment M, Mathien F, Bally P, Yangzom D, Li JZ, Goyens P. 2012. Growth, nutritional status, and signs of rickets in 0-5- year- old children in Kashin–Beck disease endemic area of central Tibet. *European Journal of Pediatrics*. 171980: 1185-91.

Singh, SP. 1980. Growth pattern of children of Tibetan origin living at moderate altitude, India. *Z. Morph. Anthrop* 7(2): 187-95.

Smith, C. 1999. Blood pressure of Sherpa men in modernizing Nepal. *American Journal of Human Biology* 11: 469-79.

Smith, C. 1997. The Effect of Maternal Nutritional Variables on Birth Weight Outcome of Infants Born to Sherpa Women at Low and High Altitude Nepal. *Hum. Biol.*9:751-763.

Subba, TB. 1990. Flith and Adaptation. Tibetan Refugees in Darjeeling- Sikkim Himalaya. Oxford University Press, New Delhi.

Torroni, A, Miller JA, Moore LG, Zamudio S, Zhuang J, Droma T, Wallace DC. 1994. Mitochondrial DNA analysis in Tibet: implications for the origin of the Tibetan population and its adaptation to high altitude. *American Journal of Physical Anthropology* 93: 189-199.

Tripathy, V, Gupta R. 2005. Birth weight among Tibetans at different altitudes in India: are Tibetans better protected from IUGR? *American Journal of Human Biology* 17:442-450.

Tripathy, V, Satapathy KC, Gupta R. 2006. ABO and Rh Polymorphism among Tibetans in India. *Human Biology*. 78(2).

Tripathy, V, Satapathy KC, Gupta R. 2006. Nutritional status and hypertension among Tibetan adults in India. *Journal of Human Ecology*. 14:77-82.

Tripathy, V, Gupta R. 2007. Growth Among Tibetans at High and Low Altitude in India. *American Journal of Human Biology.* 19:789-800.

Tripathy, V, Gupta R. 2005. Birth weight among Tibetans at Different Altitude in India: Are Tibetans Better Protected from IUGR? *American Journal of Human Biology.* 17(4): 442-50.

Satapathy, KC. 2009. Coping with Malaria: A Study of Knowledge, Attitude, and Practice (KAP) on Malaria among the Tibetans of Chandragiri, Odisha. *Mekal Insight,* I(1):68-87.

Satapathy, KC. 2010. Birth weight of Tibetans in Chandragiri, Odisha. *South Asian Anthropologist.*

Satapathy, KC. 2010. Refugees' Health: A Bioanthropological Study on Tibetan Refugees of Chandragiri, Odisha, Athena Books. Bhubaneswar.

Setty, PS and James WPT. 1994. Body mass index- A measure of chronic energy deficiency in adults. FAO Food and Nutrition paper 56. Food and Agriculture Organisation of the United Nations, Rome.

Sun, S. 1986. Epidemiology of hypertension on the Tibetan Platues. *Human Biology.* 58: 507-515.

Weiner, JS and Lourie JA. 1981.Ed. Practical Human Biology. Academic Press INC. London.

Weitz, CA. 1982. Blood pressure at rest and during exercise among Sherpa and Tibetan migrants in Nepal. *Social Science and Medicine* 16: 223-231.

Weitz, CA. 1984. Is Sherpa Response to High Altitude Unique. In: Proceedings of the Indian Statistical Institute Golden Jubilee International Conference on human Genetics and Adaptation. Vol. 2, pp. 33-49. Ed. Basu A, and Malhotra KC. *Indian Statistical Institute*, Calcutta.

Weitz, CA and Garruto RM. 2004. Growth of Han migrants at high altitude in central Asia. *American Journal of Human Biology* 16:405-19.

Weitz, CA, Garruto RM, Chin C, Liu J, Liu R, He X. 2002. Lung function of Han Chinese born and raised near sea level and high altitude in Western China. *American Journal of Physical Anthropology* 14:494-10.

Weiwen, H. 2001. Qinghai-Tibet Plateau: Human Occupation in the Early Periods. China's Tibet 12(2):1-3. (Http://www.zytzb.org.cn/xizhang/ctibet200102 (Accessed on 10/21/2002).

World Health Organisation. 1984. The incidence of low birth weight: An update. Weekly Epidemiological Record 59:205-11. (Cited by Mondal B, 1997).

Zamudio, S, Droma T, Norkyel KY, Acharya G, Zamudio JA, Niermeyer SN, Moore LG. 1993. Protection from intrauterine growth retardation in Tibetans at high altitude. *American Journal of Physical Anthropology* 91:251-64.

5

Growth Pattern of the Adolescent: A Cross-sectional Study among the Bhotia of Uttarakhand, India

—Chittaranjan Mandal

Abstract

In Uttaranchal, the Bhotias are one of the earliest inhabitants of the Kumaon and Garhwal hills. The Bhotias are divided into eight sub-groups on the basis of religion, territory, occupation and dialects. The present data were collected among the Tolcha and Marcha sub-groups of the Bhotia residing in Chamoli district of Uttarakhand. The Bhotia households in this area are located in 6,107 ft. above mean sea level. The sample consists of 282 subjects. It is interesting to note that Bhotias in this altitude depend upon different kind of food habit. Findings of the present study were compared with other published reports. It can be concluded from the comparative analysis the Bhotias; Garhwali Rajputs and Garhwali Brahmins exhibit some differences in their pattern of growth. Further, these differences may be attributed to genetic variation, differences in demographic parameters and due to differences of altitudinal zone.

Key word: *Stature, weight, BMI, Altitude, Himalayan ecology.*

Introduction

A good number of world population lives at altitudes ranging from 3000 to 5200 m above sea level but it has not been established yet whether altitude constitutes a variable affecting energy requirements (Picon-Reategui, 1982). Dietary and lifestyle changes have profound influences on the body mass and nutritional status of individuals living in developed countries. From a study of the prevalence of obesity among the Sherpa women (Smith, 1998), the mean body weight and BMI were significantly higher for the urban low altitude sample (53.3 kg) than for the total high altitude sample (51.4 kg). The rural, high altitude sample had the lowest mean body weight (50.1 kg), triceps skinfold (18.5 mm), and BMI (21.3 kg).

Mitchell and Edman (1951) are of the opinion that "there seems to be little reason to expect, and scanty evidence to support, a belief that altitude and its associated low oxygen tension will disturb the nutrient requirements of man. The low oxygen supply at high altitude will depress rather than stimulate processes, and, even more surely than a hot humid environment, will discourage the performance of muscular work and lower the capacity of work". Picon-Reategui (1982) further said that hypoxia has a stressful effect on physiology of the animal organism. However, other components of the high altitude environment, such as, cold, rough terrain and low vapour tension, may also affect metabolic processes and, of course, nutrient requirements. Nutritional status and prevalence of hypertension and the relationship between them in adult Tibetan refugee population in India at three different places have been investigated by Tripathy *et al.*, (2006). No relationship between nutritional status and body composition with blood pressure was found by them. In terms of nutritional status and hypertension Tibetans in India resemble populations from developed countries who are described as urban having a sedentary way of life.

Area and People

The Bhotia boys of Chomoli district of Uttarakhand were selected for the present study. The Bhotias are a tribal population, with Mongoloid morphological features who speak Tibeto-Burman family of languages. The word Bhotia is a generic term for several groups of people inhabiting the ranges along the snowy peaks of the Himalayas. Etymologically, the word Bhotia is popularly believed to have originated from the term *bhot* or more correctly *Bod* which means Tibet (Walton, 1911).

In Uttarakhand, the Bhotia are one of the earliest inhabitants of the Kumaon and Garhwal hills. They consume large quantities of rice, wheat and pulses. Consumption of meat, milk and milk products is low. The Bhotia are divided into eight Sub-Groups on the basis of religious practices, Territory, occupation and dialect. The Sub-Groups are: 1. Jad, 2. Tolchha, 3. Marchha, 4. Johari, 5. Jethora, 6. Darmi, 7. Chaudansi and 8. Byansi. The Bhotia practice endogamy at the territorial level and exogamy at the clan/lineage level. The present data were collected on 'Tochha' and 'Marchha' in and around towns of Joshimath and Gopeshwar of district Chamoli in Uttarakhand.

The children for the present study invariably are from families of agriculturist, and belong to nuclear families. Majority of the students come from medium size family (i.e., 4-6 members). In most of the cases (61.70 per cent), family income is invariably between Rs 1000 Rs 4999 p.m. (i.e., lower income group). 17.38 per cent of fathers were illiterate and 8.86 per cent of father had graduate level of education. On the other hand 39.36 per cent of the subjects have illiterate mothers also.

Methods

Data for the present study were collected from 282 males, ranging in age from 11+ to 17+ years, by cross-sectional method. Measurements were obtained in accordance with the techniques recommended by Weiner and Lourie, 1981. The study area is located 6000 ft. above the sea level. Date of birth of each subject was obtained from school registers/birth certificates. Age was calculated as per decimal calendar. Parent's occupation, income, education etc. were obtained from school records and enquiring from subjects. Maturational indicators like adult voice, facial hair, axial hair and pubic hair were recorded by status quo method. Subjects were asked about presence or absence of the indicators. It is rural area. Statistical analysis was performed with the help of SPSS 12.0 for windows.

Results

Descriptive statistics of age of samples recruited for present investigation is evident from Table 5.1. It is apparent that the average age in all age groups studied were nears the mid-point of age groups.

Standard deviation values range from 0.27 to 0.32, similarly standard error varies from 0.04 to 0.05.

Growth pattern of different anthropometric characteristics are presented in Table 5.2, while distance curves of different measurements are presented in Figure 5.1 and velocity curves are shown in Figure 5.2. It is apparent that body weight increased with the advancing age. An average of 27.21 kg. of weight was gained in six years from 11-17 years. However, maximum increase in weight was attained at 14 years (6.37 kg.), thereby showing adolescent spurt between age groups 13-14 years, followed by a relatively low increment at 16 years (3.44 kg.) and 17 years (3.36 kg) age groups. Like body weight, stature also increased gradually with the advancing age. Initially rapid increment of height was gained between age groups 11-12 (6.57 per cent) and followed by 13-14 (6.56 per cent) years. On the other hand relatively low increment of height was gained between age groups 16 to 17 years (1.43 per cent). Adolescent spurt occurred between 13-14 years with per annum change of 9.55 cm.

Table 5.1. Distribution of boys according to different age groups

Age Groups	*No.*	*Mean*	*S.D.*	*SE*	*Minimum*	*Maximum*
11	40	11.43	0.28	0.05	11.02	11.99
12	40	12.47	0.29	0.05	12.03	12.99
13	40	13.51	0.28	0.04	13.08	13.97
14	41	14.44	0.32	0.05	14.01	14.96
15	40	15.56	0.277	0.04	15.05	15.96
16	40	16.55	0.31	0.05	16.04	16.88
17	41	17.57	0.28	0.04	17.01	17.95

The Table clearly shows that the sitting height gradually increased with advancement of age from 69.51 cm at 11 year age group to 86.63 cm at 17 years with a total change of 17.12 cm. Relatively maximum increment is observed between ages 13 to 14 years (5.74 per cent), followed by 11 to 12 years (5.29 per cent). Age group 16-17 years marked relatively slow growth with an increment of 1.31 per cent. However; adolescent spurt was clearly discernible between age groups 13 to 14 years with an increment of 4.41 cm (5.74 per cent).

A clear-cut increase in total arm length is observed with increase in age from 55.88 cm at 11 years to 70.92 cm at 17 years of age, showing a gain of 15.04 cm. in total arm length. Maximum increase is attained at 14 years (94.71 cm; 7.54 per cent), thereby showing adolescent spurt between age groups 13-14 years. Next highest increment is recorded between age groups 11-12 years (6.91 per cent) followed by (12-13) years (4.58 per cent). Whereas, lowest increment is noted between the age groups 16 to 17 years (1.24 per cent).

Leg length increased with the advancing age. An average of 16.24 cm. of length is gained in six years. However, maximum increase in leg length is attained at 14 years (5.14 cm), thereby showing adolescent spurt between age groups 13-14 years. Highest percentage of absolute growth is observed between age groups 11 to 12 years. Relatively low increments have been found in age groups 15 to 16 years (1.52 per cent) and 16 to 17 years 1.56 per cent.

Table 5.2. Mean, S.D. and per annum growth (in per cent) for different measurements

MEASUREMENT		AGE GROUP						
		11	12	13	14	15	16	17
Weight	MEAN	26.09	31.50	35.17	41.54	46.50	49.94	53.30
	S.D	3.40	5.61	5.97	6.07	7.10	7.00	5.61
	PAG (per cent)		20.74	11.65	18.11	11.94	7.40	6.73
Height	MEAN	130.11	138.66	145.57	155.12	158.84	161.87	164.19
	S.D.	5.54	6.45	7.77	6.44	8.65	5.54	4.86
	PAG (per cent)		6.57	4.98	6.56	2.40	1.91	1.43
Sitting height	MEAN	69.51	73.19	76.84	81.25	83.61	85.51	86.63
	S.D.	2.33	3.60	4.24	4.30	4.44	3.53	3.00
	PAG (per cent)		5.29	4.99	5.74	2.90	2.27	1.31
Total arm length	MEAN	55.88	59.75	62.49	67.2	68.69	70.05	70.92
	S.D.	2.79	3.26	3.55	2.68	3.91	2.69	2.46
	PAG (per cent)		6.92	4.58	7.54	2.22	1.97	1.24
Leg length	MEAN	60.61	65.46	68.73	73.87	75.22	76.36	77.55
	S.D.	4.06	3.83	4.51	3.00	4.78	3.42	3.90
	PAG (per cent)		8.00	5.99	7.48	1.83	1.51	1.56
Biacromial diameter	MEAN	28.51	30.56	32.03	34.02	35.90	36.67	77.55
	S.D.	1.33	1.52	1.64	1.58	1.83	1.57	1.56
	PAG (per cent)		7.19	4.81	6.21	5.53	2.14	1.39
Bi-illiocristal diameter	MEAN	19.82	20.80	21.76	22.97	23.97	24.18	25.05
	S.D.	1.08	1.58	1.64	1.58	1.83	1.57	1.56
	PAG (per cent)		4.94	4.61	5.56	4.35	0.88	3.60

(Contd...)

Neck circumference	MEAN	26.26	27.60	28.37	30.08	31.88	32.55	33.31
	S.D.	1.00	1.48	1.67	1.84	1.76	1.95	1.42
	PAG (per cent)		5.10	2.79	6.03	5.98	2.10	2.33
Chest circumference	MEAN	61.88	66.35	68.52	73.17	77.52	79.45	81.76
	S.D.	3.48	5.23	4.59	5.84	5.21	5.29	4.13
	PAG (per cent)		7.22	3.27	6.78	5.94	2.49	2.91
Mid-upper Arm Circumference	MEAN	16.95	18.58	19.22	20.63	22.28	23.00	23.88
	S.D.	1.48	2.07	1.93	1.93	2.62	1.88	1.80
	PAG (per cent)		9.62	3.44	7.34	8.00	3.23	3.83
Calfcircumference	MEAN	24.74	26.99	27.96	29.74	31.177	31.76	32.10
	S.D.	1.58	2.18	2.21	2.10	2.14	2.19	1.81
	PAG (per cent)		9.09	3.59	6.37	4.81	1.89	1.07
Tricep skin fold	MEAN	5.79	6.14	5.97	5.48	5.33	5.40	5.85
	S.D.	1.71	2.63	2.01	1.59	1.29	1.66	2.16
	PAG (per cent)		6.04	-2.77	-8.21	-2.74	1.31	8.33

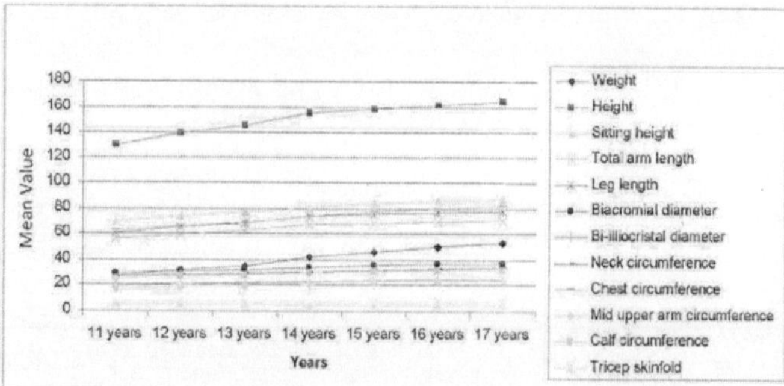

Figure 5.1: Distance curve.

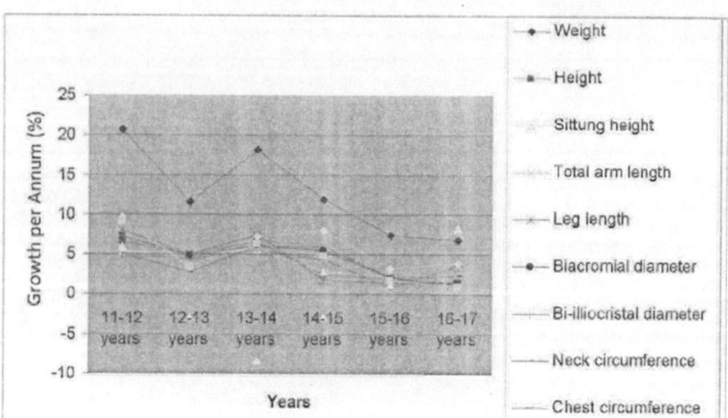

Figure 5.2: Velocity curve.

Bi-acromial diameter gives the maximum shoulder width of the subject. It is observed from the table that the mean value of Bhotia boys increased gradually from age group 11 to 17 years recording a total increment of 8.67 cm. Maximum increase in bi-acromial diameter is attained at 12 years (2.05 cm), therefore, adolescent spurt occurred between age groups 11 to 12 years. Significant per annum growth occurred upto 15 years of age.

Bi-illiocristal diameter gives maximum width of the pelvis. Like shoulder width, Bi-illiocristal diameter recorded gradual increment in average values upto age group 17 years, with a total increment of 5.23 cm. Maximum increase in bio-illiocristal diameter is attained at 14 years (1.21 cm; 5.56 per cent), thereby showing adolescent spurt between age groups 13 to 14 years.

It revealed that neck circumference increased gradually from age 11 to 17 years recording a total increment of 7.05 cm. Maximum per annum growth is recorded between age groups 14 to 15 years (1.80 cm), showing adolescent spurt. Chest circumference recorded a total average gain of 19.88 cm with gradual increase in chest dimensions with growing age. Maximum increment is recorded at age group 14 years (44.65 cm.), thereby reflecting adolescent spurt between age groups 13 to 14 years. However, significant interage group changes are recorded from 11 to 15 years and thereafter 16 to 17 years.

It is observed from the Table 5.2 that mid-upper arm circumference increased with growing age with a total average increment of 6.93 cm from 11 to 17 years. Maximum increment is recorded between age groups 14 to 15 years (1.65 cm); Alike other parameters, calf circumference increased with increment in age, with a total average increment of 7.36 cm. Maximum increase in calf circumference is attained at 12 years (2.25 cm.), thereby showing adolescent spurt between age groups 11 to 12 years. Thereafter, though the average measurements increased, but tempo of increment fluctuated.

The mean values of Tricep Skinfold thickness fluctuated with growing age. It recorded slight increment between age groups 11 to 12 years. For next three years; a decrease in skinfold measurements is found, which is more marked between the age groups 13-14 years. The skinfold thickness at tricep again showed increase in following years i.e., from 16 to 17 years with maximum per annum increment between age groups 16-17 years (8.33 per cent).

Table 5.3: Age wise mean BMI and SD among Bhotia boys

Age group in year	*Frequency*	*Mean BMI*	*SD*
11	40	15.36	1.24
12	40	16.32	2.22
13	40	16.49	1.57
14	41	17.18	1.59
15	40	18.33	1.72
16	40	19.00	2.11
17	41	19.76	1.75
All ages	**282**	**17.50**	**2.29**

Table 5.3 clearly shows that BMI values is increasing with the advancement of age and mean BMI value for the total population is 17.50 Kg/m^2. On the basis of BMI, 69.50 per cent of boys were found underweight, 28.01 were normal, whereas 2.4 per cent were found overweight and obese.

Discussion

It can be seen from various observations that there is often a variation in the tempo of growth in different body-parts. Externally this means that human body is continuously changing its size. The variation of the speed of growth between one body part and another is regulated, so that each part reaches its size, proportional to the role, it has to play in the body's physiology at the correct time.

The analysis of different measurements clearly point to the adolescent spurt at age groups 13-14 years in most of the measurements. Bi-acromial diameter and calf circumference attain peak velocity between age groups 11-12 years. Tricep skinfold thickness decreased during age group 12-13 years, and gradually increase to attain peak values at age of 17 years. Values of Body Mass Index increase with advancing age, which record peak increments between age groups 14-15 years.

Table 5.4: Comparison of growth among Bhotia and Garhwali boys

MEASUREMENT		AGE													
		11		*12*		*13*		*14*		*15*		*16*		*17*	
		Bhotia	***Garhwali Rajput***	***Bhotia***	***Garhwali Rajput***	***Bhotia***	***Garhwali Rajput***	***Bhotia***	***Garhwali Rajput***	***Bhotia***	***Garhwali Rajput***	***Bhotia***	***Garhwali Rajput***	***Bhotia***	***Garhwali Rajput***
Weight	N	40	42	40	46	40	49	41	47	40	45	40	47	41	40
	Mean	26.09	31.26	31.50	35.13	35.17	39.60	41.54	42.70	46.50	47.60	49.94	49.06	53.30	50.86
	SE	0.54	0.85	0.89	0.82	0.94	1.04	0.95	1.05	1.12	1.16	1.11	1.17	0.88	1.20
Height	N	40	42	40	46	40	49	41	47	40	45	40	47	41	40
	Mean	13.11	139.61	138.66	146.48	145.57	154.91	155.12	158.89	158.84	165.70	161.87	164.87	164.19	165.89
	SE	0.88	1.21	1.02	1.40	1.23	1.32	1.01	1.22	1.37	1.24	0.88	1.02	0.76	0.97
Sitting height	N	40	42	40	46	40	49	41	47	40	45	40	47	41	40
	Mean	13.11	139.61	138.66	146.48	145.57	154.91	155.12	158.89	158.84	165.70	161.87	164.87	164.19	165.89
	SE	0.37	0.56	0.57	0.79	0.67	0.79	0.67	0.68	0.70	0.66	0.56	0.60	0.47	0.49
Total arm length	N	40	42	40	46	40	49	41	47	40	45	40	47	41	40
	Mean	55.88	59.90	59.75	63.15	62.49	67.83	67.20	69.02	68.69	71.80	70.05	71.57	70.92	72.84
	SE	0.44	0.57	0.52	0.68	0.56	0.72	0.42	0.62	0.62	0.56	0.43	0.53	0.38	0.62
Leg length	N	40	42	40	46	40	49	41	47	40	45	40	47	41	40
	Mean	60.611	67.17	65.46	70.60	68.73	75.81	73.87	76.92	75.22	80.14	76.36	79.50	77.55	79.20
	SE	0.64	0.76	0.61	1.00	0.71	0.74	0.47	0.70	0.76	0.72	0.54	0.75	0.61	0.68
Biecromial diamaeter	N	40	42	40	46	40	49	41	47	40	45	40	47	41	40
	Mean	28.51	30.63	30.56	32.11	32.03	33.67	34.02	34.69	35.90	36.50	36.67	36..81	37.18	37.63
	SE	0.21	0.40	0.24	0.34	0.33	0.40	0.45	0.37	0.41	0.41	0.35	0.34	0.25	0.40
Bi-illiocristal diameter	N	40	42	40	46	40	49	41	47	40	45	40	47	41	40
	Mean	19.82	22.22	20.80	23.40	21.76	24.62	22.97	25.08	23.97	26.30	24.18	26.22	25.05	26.65
	SE	0.17	0.23	0.25	0.22	0.26	0.48	0.25	0.26	0.29	0.32	0.25	0.26	0.24	0.24

(Contd...)

MEASUREMENT		AGE													
		11		12		13		14		15		16		17	
		Bhotia	Garhwali Rajput	Bhotia	Garhwali Rajput	Bhotia	Garhwali Rajput	Bhotia	Garhwali Rajput	Bhotia	Garhwali Rajput	Bhotia	Garhwali Rajput	Bhotia	Garhwali Rajput
Neck circumference	N	40	42	40	46	40	49	41	47	40	45	40	47	41	40
	Mean	26.26	27.36	27.60	27.47	28.37	29.07	30.08	30.20	31.88	31.20	32.55	31.81	33.31	32.35
	SE	0.16	0.29	0.23	0.39	0.26	0.25	0.29	0.29	0.28	0.30	0.31	0.30	0.22	0.31
Chest circumference	N	40	42	40	46	40	49	41	47	40	45	40	47	41	40
	Mean	61.88	63.75	66.35	66.31	68.52	69.36	73.17	71.83	77.52	74.43	79.45	75.91	81.76	76.69
	SE	0.55	0.61	0.83	0.70	0.73	0.63	0.91	0.71	0.82	0.78	0.84	0.86	0.65	0.80
MUAC	N	40	42	40	46	40	49	41	47	40	45	40	47	41	40
	Mean	16.95	17.86	18.58	19.02	19.22	19.47	20.63	20.10	22.28	21.20	23.00	21.58	23.88	21.99
	SE	0.24	0.24	0.33	0.31	0.31	0.26	0.30	0.29	0.41	0.31	0.30	0.34	0.28	0.30
Calf circumference	N	40	42	40	46	40	49	41	47	40	45	40	47	41	40
	Mean	24.74	26.29	26.99	27.93	27.96	29.21	29.74	30.13	31.17	30.74	31.76	30.91	32.10	31.02
	SE	0.25	0.35	0.35	0.35	0.35	0.33	0.33	0.34	0.34	0.30	0.35	0.40	0.28	0.45
Tricep skinfold	N	40	42	40	46	40	49	41	47	40	45	40	47	41	40
	Mean	5.79	7.61	6.14	7.66	5.97	6.89	5.48	6.88	5.33	7.02	5.40	7.05	5.85	7.78
	SE	0.27	0.28	0.42	0.49	0.32	0.30	0.25	0.35	0.20	0.37	0.26	0.51	0.34	0.54
BMI	N	40	42	40	46	40	49	41	47	40	45	40	47	41	40
	Mean	15.36	15.91	16.32	16.30	16.49	16.40	17.18	16.80	18.33	17.28	19.00	17.94	19.76	18.42
	SE	0.20	0.26	0.35	0.24	0.25	0.25	0.25	0.30	0.27	0.33	0.33	0.26	0.27	0.34

Comparison of all anthropometric parameters of the present study with the Garhwali Rajput Boys (Kaul and Tandon, 2001: Table 5.4) studied in Dehradun show that in case of Height, total arm length, Leg length, Bi-acromial breadth, Bi-cristal breadth and Tricep Skinfold, the Garhwali Rajput boys have greater dimensions than Bhotia boys at all age groups. Same trend is found for sitting height, except at 16 years, where Bhotias shows greater measurements. Weight and calf circumference are greater among the Garhwali Rajputs from 11-14 years, whereas mid upper arm circumference show greater dimensions upto 13 years, and thereafter, Bhotia attain higher mean values. Neck and chest circumference are greater in Bhotias, except at 13 years for both measurements and in addition at 14 years for neck circumference. Except at 11 years, Bhotia have higher mean values for BMI, than Garhwali Rajput at all age group, thereby showing proportionately higher weight. The adolescent spurt occurred between the ages 13-14 years among Bhotias in most of the measurements, which is one year prior to that found among Garhwali Rajputs (i.e., 14-15 years).

The growth data have also been compared with NCHS Standards (1966-70), ICMR Standards (1972), ICMR U.P. data (1972), based on data collected during 1956-1966 and Garhwali Brahmin populations of Dehradun (Tandon and Kaul, 2003) in Table 5.5 and depicted in Figures 5.3 and 5.4. The Bhotia boys are short in stature and lighter in body weight as compared to the NCHS reference data. At the same time they are heavier and taller than ICMR reference data, except in the weight of ICMR U.P. data and in the height of ICMR standard and ICMR U.P. data at 11 years age. Upto 14 years of age, Garhwali Brahmins are heavier than the Bhotia, thereafter, the Bhotia outweight the Garhwali Brahmins. So far height is concerned, the Garhwali Brahmins are taller than Bhotias at all age groups.

Table 5.5: Comparison of weight and height among different populations

Age	*Bhotia*	*NCHS 1966-70*	*ICMR*	*ICMR U.P.*	*GARHWALI RAJPUTS*	*GARHWALI BRAHMINS*
			Weight (in Kgs)			
11	26.09	38.4	25.9	26.3	31.26	31.22
12	31.5	42.98	28.5	28.8	35.13	34.48
13	35.17	50	32.1	32.6	39.6	38.93
14	41.54	56.66	35.7	36.6	42.7	42.81
15	46.5	61.61	39.6	41	47.6	45.33
16	49.94	64.82	43.2	44.3	49.06	49.7
17	53.3	68.03	45.7	46.2	50.86	51.05
			Height (in cms)			
11	130.11	145.7	133.4	132.8	139.61	140.37
12	138.66	152.3	138.3	137.8	146.48	147.91
13	145.57	159.8	144.6	145.1	154.91	154.28
14	155.12	166.7	150.1	150.8	158.89	160.15
15	158.84	171.4	155.5	157.1	165.7	163.95
16	161.87	174.3	159.5	160	164.87	166.51
17	164.19	175.5	161.4	160.6	165.89	166.79

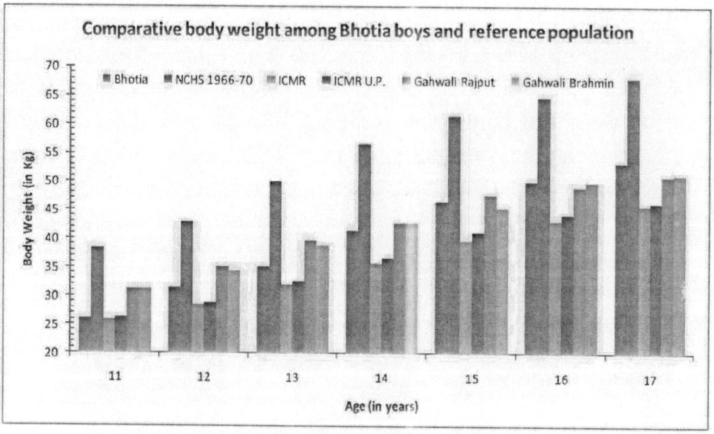

Figure 5.3: Compartive body weight among Bhotia boys and reference population.

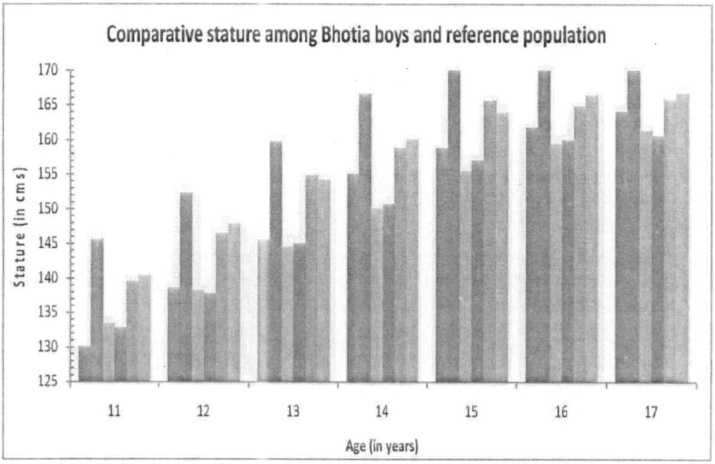

Figure 5.4: Compartive stature among Bhotia boys and reference data.

It can be inferred from the comparative analysis of Bhotias and other populations that Bhotias, Garhwali Rajputs and Garhwali Brahmins exhibit some differences in their pattern of growth. It may further be stated that these differences, may be attributed to genetic variation, differences in demographic parameters and due to differences of altitudinal zone. All these points were also supported by the previous studies (Frisancho and Baker, 1970; Malhotra, 1989; Singh and Surinder Nath, 1997). The well off conditions of the present sample than ICMR standards and ICMR U.P. data, can be explained due to comparatively good environmental conditions, medium size family, besides increased interventional programmes being provided by government.

Acknowledgements

This paper is based on the project entitled, "The Physical Growth of Adolescent", which was approved by the Anthropological Survey of India, Kolkata. The author is grateful to the said Organisation. The author is also grateful to all the Bhotia participants who participated in this study.

References

Baker, P.T. 1978. *The biology of high altitude peoples.* Cambridge University Press, Cambridge.

Bhattacharya, K.K., S. Mandal and, G.C. Ghosh. 1994. Nutritional Status of the Oraon of Sundarban, West Bengal. *J. Anthrop. Survey India* 43: 31-36.

Ferro-Luzzi, A., S. Sette, M. Franklin and W.P.T. James 1992. A simplified approach of assessing adult chronic energy deficiency. *Eur. J. Clin. Nutrition* 46: 173-186.

Frisancho, A.R. and P.T. Baker 1970. Altitude and growth, a study of patterns of physical growth of a high altitude Peruvian Quechua population. *American Journal of Physical Anthropology*. 32:272-92.

Indian Council of Medical Research (ICMR) 1972. Growth and physical development of Indian infants and children. Technical Report Series No. 18. ICMR, New Delhi.

James, W.P.T., A, Ferro-Luzzi, and J.C. Waterlow 1988. Definition of chronic energy deficiency in adults. *Eur. J. Clin. Nutrition* 42: 969-981.

Kaplan, N.M. and L.H. Opie 2006. Controversies in hypertension. *Lancet* 367: 168-176.

Kaul, V. and V.K.Tandon 2003. Adolescent Growth of Garhwali Rajput Boys (Report), submitted to the Anthropological Survey of India, Kolkata.

Malhotra, R 1989. *Bio Anthropology of High Altitude People*, Mittal Publications, New Delhi.

Marshall, William and J.M.Tanner 1986. *Puberty. In Human Growth* Vol. 2. Ed.Frank Falkner & J.M. Tanner. Plenum Press, N.Y.

Martin and Saller cited from Singh, I.P. and M.K. Bhasin 1968. *Anthropometry*, Kamala Raj Enterprises, Delhi.

National Center For Health Statistics 1970. Height and weight of children- United States. Series No. 11, No. 104 (Children aged 6 years to 11 years) U.S. Department of Health, Education and Welfare Public Health Services.

Park, K. 2002. *Text book of preventive and social medicine*. Benarasideas Bhanot, Jabalpur.

Picon-Reategui, E. 1978. The food and nutrition of high altitude populations.- In: Baker, P.T. (ed.): *The biology of high altitude peoples*.- Cambridge University Press, Cambridge.

Rolland-Cachera, M.F. 1993. Body composition during adolescence : Methods, limitations and determinants. *Hormone Research* 39: (Suppl. 3), 25-40.

Shetty, P.S. and James, W.P.T. 1994. Body Mass Index : A Measure of Chronic Energy Deficiency in Adults. *FAO Food and Nutrition*, Paper 56, Rome.

Singh, Ravinder and Surinder Nath 1997. Comparative Growth Patterns among the Jatav and the Chauhan males of district Moradabad, *U.P. Journal of Anthropological Survey of India* 46(1) : 69-77.

Smith, C. 1998. Prevalence of Obesity and Contributing factors among Sherpa Women in Urban and Rural Nepal. *Am. J. Hum. Biol.* 10:519-528.

Stamler, J. 1991. Blood pressure and high blood pressure: Aspects of risk. *Hypertension* 18: 95-107.

Walton cited from Singh, K.S. 1911, 1994. People of India. The Schedule Tribes. National Series Volume 111. Anthropological Survey of India, Kolkata, Oxford University Press. Delhi.

WHO 1995, The use and interpretation of anthropometry. Report of WHO expert committee. WHO Technical report Series 854.

Theme – 2
Under-Nutrition and Disease

6

Prevalence of Thinness, Overweight and Obesity among Karbi Tribal Children Aged 5-12 Years of Karbi Anglong, Assam Northeast India

—Nitish Mondal, Rongkehai Ronchehon and Nitamoni Bharali

Abstract

The prevalence of malnutrition among children is an important indicator of health status of a community or population. It is one of the significant areas of concern in the developing countries where a vast proportion of the population is malnourished. The aims and objectives of the study are to determine and compare the double burden of malnutrition using the recommended anthropometric measures of thinness, overweight and obesity among Karbi tribal children of Assam, India. The present cross-sectional study has been carried out among the 480 (239 boys; 241 girls) Karbi tribal school going children aged 5 to 12 years of Karbi-Anglong district of Assam, Northeast India, using multistage stratified random sampling method. The anthropometric measurements of height and weight were recorded using standard anthropometric procedures. The prevalence of thinness, overweight and obesity was assessed using the recently proposed international reference and cut-offs proposed by Cole et al., The data was analysed using the help of Statistical Package of Social Science (Version, 17.0). The prevalence of overall thinness was found to be 17.15 per cent and

20.75 per cent among boys and girls ($p>0.05$). The prevalence of overall combined overweight-obesity (BMI ≥ 25.00 kg/m^2) was found to be 17.15 per cent and 11.20 per cent among boys and girls. The overall prevalence of overweight (17.15 per cent vs. 11.20 per cent) and obesity (3.76 per cent vs. 2.90 per cent) was found to be greater among the boys than the girls ($p>0.05$). The prevalence of higher degree of malnourishment is transversely persistent among the Indian children. The appropriate intervention should be introduced to improve the nutritional status.

Keywords: *Overweight, obesity, Thinness, Malnutrition, Anthropometry, India.*

Introduction

The prevalence of malnutrition among children is an important indicator of health status of a community or population. It is one of the significant areas of concern in many of the developing countries where a vast proportion of the population is malnourished (Nandy *et al.*, 2005; Svedberg, 2011). It has been estimated that approximately 70.00 per cent of the world's malnourished children live in Asia, and 800 million individuals being reportedly undernourished worldwide, a little under one-third (258 million individuals) are concentrated in South Asia (Ahmed *et al.*, 2012). The prevalence of malnutrition is one of the principal causes of ill-health condition and leads to causes premature mortality and morbidity among children of the developing countries such includes India (Nandy *et al.*, 2005; Svedberg, 2011). A large scale population investigation has shown that the prevalence of stunting (low-height-for-age) was observed to be around 50.0 per cent among children aged below five years in 2005-06 reported in National Family Health Survey (NFHS-3), reflecting a long-term chronic undernutrition, while the proportion of underweight (43.0 per cent) was almost double that reported in sub-Saharan Africa (IIPS and Macro International, 2007). Therefore, the assessment of nutritional status among the vulnerable segments of the populations, thus assumes an immediate priority and is a daunting task for any researcher. The prevalence of childhood under-nutrition is considered to have significant detrimental health effects in those children who survive to adulthood (WHO, 1995; Svedberg, 2011). Studies on nutritional assessment among children in India continue to be very challenging task, where a large proportion of the population is generally poverty-stricken population groups, socio-economic disparities, high percentage of illiteracy and inadequate access to health facilities all contributes serious challenges to such studies. Ironically, despite economic development, India continues to have a high prevalence of infant and childhood mortality and malnutrition (Khor, 2008; Bamji, *et al.*, 2008; Svedberg, 2011; Ramachandran and Gopalan, 2011; Kapil and Sachdev, 2012; Varadharajan *et al.*, 2013). Several researchers have already advocated that an urgent efforts are needed to reduce this mutually reinforcing linkage between poverty and under-nutrition (Ramachandran, 2007; Svedberg, 2011; Varadharajan *et al.*, 2013).

The prevalence of childhood under-nutrition appears to have slightly decreased globally except in Africa, and the absolute number of undernourished children has increased because of the overall population growth. In India, the legacy of child malnutrition among preschool children is a major public health problem and an impending obstacle to overall national and economic development in India (Khor, 2008; Varadharajan *et al.*, 2013). Several researchers have already reported that the prevalence of under-nutrition is a major public health problem among Indian children (Nandy *et al.*, 2005; Bose *et al.*, 2007; Seetharaman *et al.*, 2007; Mandal and Bose 2009; Mondal and Sen, 2010a,b; Ramachandran and Gopalan, 2011; Sen *et al.*, 2011; Sen and Mondal, 2012; Rao *et al.*, 2012; Prasad *et al.*, 2012; Mondal, 2014; Tigga *et al.*, 2015). A significant number of studies have reported that socio-economic, demographic and maternal factors are found to have a strong association with the prevalence of under-nutrition among children (Rahman and Chowdhury, 2007; Zottarelli *et al.*, 2007; Hien and Kam, 2008; Mondal and Sen, 2010b; Sen and Mondal, 2012; Martorell and Young, 2012; Tigga *et al.*, 2015). Several studies have also reported that the prevalence of under-nutrition was found to be very high among children aged 5-12 years in India (Mittal and Srivastava, 2006; Mitra *et al.*, 2007; Chowdhury *et al.*, 2008; Mondal and Sen, 2010b; Sen and Mondal, 2012).

The current nutritional investigations have shown that the prevalence of under-nutrition is the predominant form of malnutrition, but researchers have reported a significant acceleration observed in the incidences of overweight-obesity, especially among higher socio-economic groups residing in urban and suburban regions in India

(Sharma *et al.*, 2007; Sidhu *et al.*, 2006; Seth and Sharma, 2013; Wang *et al.*, 2009; Kapil and Sachdev, 2012; Varadharajan *et al.*, 2013). Nevertheless, the urban advantage has been fading in the last few decades, and since the fast changes in the diet and lifestyle of city dwellers resulting from industrialisation, urbanisation and globalisation, have triggered marked consequences for the overall health and nutritional status (Yang *et al.*, 2009; Varadharajan *et al.*, 2013). There is growing recognition of the emergence of a 'double burden' of malnutrition, with the subsequent occurrence of under- and over-nutrition among different population groups in developing countries including India (Khor, 2008; Varadharajan *et al.*, 2013; Sharma and Mondal, 2014). It is conjectured that the prevalence of excess adiposity (e.g., overweight and obesity) and it's associated with disease prevalence could become a major public health issue in the future since malnutrition during fetal development and early childhood are predisposing factors in developing countries include India. The recent trend has already shown that the prevalence of overweight and obesity are creating an extra population burden, where malnutrition and nutritional deficiencies are still persistent in young children in India. Therefore, the assessment of nutritional status of the children has very potential to play a significant role determine the prevalence of undernutrition, overweight-obesity and to evaluate the current malnutrition situation, the role and efficacy of the ongoing intervention programme and/or formulating developmental strategies related to health intervention in the populations.

Aims and objectives

The aims of the present study are briefly enumerated below:

1. To assess the physical growth pattern among the tribal Karbi children of Karbi-Anglong Assam, Northeast India.
2. To determine double burden of malnutrition using the recommended anthropometric measures of thinness, overweight and obesity among Karbi tribal children.
3. To compare the prevalence of malnutrition with different tribal and non-tribal ethnic populations of India.

Material and Method

Subjects and study area

The present cross-sectional study has been carried out among the 480 (239 boys; 241 girls) Karbi school going children aged 5 to 12 years of Karbi-Anglong district of Assam, Northeast India. The district Karbi-Anglong (25°33′ N to 26°35′ N latitude and 92°10′ E to 93°50′ E longitude) has occupied an area of 10,434 km^2 and the largest district amongst the 27 districts of Assam. The subjects were inhabitant of the Diphu Lumbajong block of Karbi-Anglong district. The community areas are mostly inhabited by the homogenous Karbi tribal population. According to the census 2011, Karbi-Anglong had a population of 9,65,280 (4,93,482 male; 4,71,798 female) with a literacy rate of 73.52 per cent (82.12 per cent male; 64.62 per cent female). Ethnically, the Karbi's are the Mongoloid ethnic group and linguistically belong to the Tibeto-Burman speaking population (Bordoloi and Thakur, 1988). The subjects were obtained using a multi-stage stratified random sampling method. A total of five government primary schools was covered during the period of December, 2012 to April 2013. The schools were selected due to their easy accessibility and homogeneity of the population using the purposive random sampling method. The age of the children was recorded from school records and subsequently verified from the birth certificates or any relevant documents. All the subjects were free from any physical deformities, systematic diseases, nutritional deficiencies and previous surgical episodes, while presences of those characteristics were excluded. The study permission was taken from the school authorities and an informed consent was also obtained from either parent of the children prior to collect the data. The present study was conducted in accordance with the ethical guidelines of human experiments as laid down in the Helsinki Declaration of 2000 (Touitou *et al.*, 2004).

Anthropometric measurements

The anthropometric measurements of height and weight were obtained using standard anthropometric procedures (Hall *et al.*, 2007). Height was recorded with the subject standing erect in a flat platform and head was oriented in the Frankfort horizontal plane from floor to vertex using anthropometer, nearest to 0.1 cm. The weight was measured wearing minimum cloth and being bare footed with the help of a portable digital weight scale nearest to 0.1 kg. Standard precautions were followed to ensure instrument's validity, and calibrations were performed from time to time. The subject covered under present study was measured with ample and precision to avoid any possible systematic errors in the process of anthropometric data collection (Harris and Smith, 2009).

The technical error of measurement ($TEM=\sqrt{(\Sigma D^2/2N)}$, D=difference between the measurements, N=number of individuals measured) was calculated using the standard method to determine the accuracy of the anthropometric measurements (Ulijaszek and Kerr, 1999). The co-efficient of reliability (R) was subsequently calculated using TEM by the following equation: $R=\{1-(TEM)^2/SD^2\}$, SD=standard deviation of the measurements. A total of 50 children was measured both by the authors (RR and NM), and the intra-observer and inter-observer TEM were observed to be within the cut-off values (R=0.95) (Ulijaszek and Kerr, 1999). Hence the measurements recorded in the present investigation were considered to be reliable and reproducible and the measurements were subsequently collected by the author RR.

Assessment of nutritional status

The prevalence of thinness was assessed using the recently proposed international reference and cut-offs proposed by Cole *et al.*, (2007). The prevalence of thinness was categories in three categories include Grade I (Mild thinness), Grade II (Moderate thinness) and Grade III (Severe thinness). This proposed cut-offs are similar to the adult cut-offs of chronic energy deficiency (CEC) (WHO, 1995). The prevalence of overweight and obesity was also assessed using the proposed reference of Cole *et al.*, (2000). These age specific cut-offs are found similar to the adult overweight (BMI $\geq$25.00 kg/m^2) and obesity (BMI $\geq$30.00 kg/m^2) (WHO, 1995).

Statistical analysis

The statistical analysis was carried out using the Statistical Package for Social Sciences (SPSS, version, 17.0). The anthropometric variables of weight, height and BMI were depicted using descriptive statistics.

Normality was tested using the Shapiro–Wilk test for anthropometric variable sex and age groups-wise, and p-values observed to be statistically not significant in most of the categories ($p>0.05$). One way analysis of variance (ANOVA) using the Scheffe procedure was done to assess sex and age-specific mean differences in anthropometric variables. The chi-square analysis was done to assess the differences in the nutritional indices between sexes in the different nutritional measures of thinness overweight and obesity. The chi-square analysis was done to assess the difference in the prevalence of malnutrition with different Indian studies. The Yates correction term was also taken into consideration terms adds to the accuracy of chi-square tests where the cells possessed less than five individuals. This correction terms adds to the accuracy of chi-square analysis determinants when the number of classes are small. A p-value of <0.05 is considered to be statistically significant.

Results

Age-and sex-specific subject distribution, means and standard deviation (±SD) of height, weight and BMI and mean differences among Karbi children are depicted in Table 6.1. The overall and age-specific mean values showed general trends that boys were found to be taller and heavier than the girl counterparts ($p<0.05$). The age specific mean values of weight and height were gradually increased with age in both boys and girls ($p<0.05$). The age specific mean BMI values are not showing any age-specific trends among boys and girls. The mean BMI was ranged from 15.27 kg/m^2 (in 5 years) to 18.75 kg/m^2 (in 10 years) among boys and 15.36 kg/m^2 (in 5 years) to 18.17 kg/m^2 (in 12 years) among girls. The mean BMI comparison with the WHO (2007) reference population showed that the majority of the Karbi children was reflecting inappropriate nutritional status and poor physical growth attainments than their normal counterparts (Figure 6.1). The majority of the boys and girls were found <75th percentiles. The mean BMI values were found to be below the 75th percentile among the girls, percentile in 12 years (<50th percentile). The sex-specific mean differences were found to be statistically significant in height, weight and BMI among boys and girls using ANOVA ($p<0.05$). Further, age-specific mean differences were found to be statistically significant in height ($F= 35.95$; d.f., 1, 238 and $F=57.18$; d.f., 1, 240), weight

(F=19.28; d.f., 1, 238 and F=29.00; d.f., 1, 240) and BMI (F=4.26; d.f., 1, 238 and F=5.44; d.f., 1, 240) among boys and girls, respectively

Table 1: Descriptive statistics (mean + SD) of collected anthropometric variables among Karbi children of Assam

Age group	*Subjects*		*Height (cm)*		*Weight (kg)*		*BMI (kg/m²)*	
	Boys	*Girls*	*Boys*	*Girls*	*Boys*	*Girls*	*Boys*	*Girls*
5 years	32	27	119.13 ±7.75	118.61 ±6.98	21.68 ±2.92	21.78 ±4.19	15.27 ±1.42	15.36 ±1.94
6 years	32	25	124.68 ±6.66	123.12 ±6.96	26.13 ±6.39	23.92 ±5.02	16.67 ±3.29	15.67 ±2.52
7 years	27	31	128.38 ±4.31	125.23 ±6.32	28.25 ±3.97	25.62 ±5.87	17.09 ±1.89	16.15 ±2.67
8 years	36	27	130.91 ±7.92	126.73 ±6.18	28.17 ±5.95	26.81 ±5.30	16.32 ±2.30	16.61 ±2.45
9 years	27	33	132.99 ±5.69	132.08 ±5.09	29.74 ±5.64	27.85 ±4.22	16.72 ±2.43	15.95 ±2.13
10 years	25	38	137.94 ±7.43	133.19 ±5.88	35.76 ±7.25	27.92 ±4.71	18.75 ±3.41	15.65 ±1.75
11 years	33	24	138.41 ±9.18	137.25 ±6.33	33.61 ±8.45	33.13 ±6.30	17.34 ±2.85	17.59 ±3.26
12 years	27	36	143.86 ±7.74	144.38 ±5.25	36.06 ±6.77	38.11 ±6.95	17.34 ±2.46	18.17 ±2.52
F-value	239	241	35.95**	57.18**	19.28**	29.00**	4.26**	5.44**

**$p<0.01$

($p<0.01$) (Table 1).

Prevalence of thinness (low-BMI-for-age)

The prevalence of overall thinness was found to be 17.15 per cent and 20.75 per cent among boys and girls (Figure 6.3). The age-sex specific prevalence of thinness was ranged from 7.41 per cent (in 7 years) to 25.00 per cent (in 5 years) and 8.33 per cent (12 years) to 29.63 per cent (5 years) among boys and girls, respectively. Age and sex-specific prevalence of different grades of thinness (low-BMI-for-age) among the tribal Karbi children is depicted in Table 6.2.

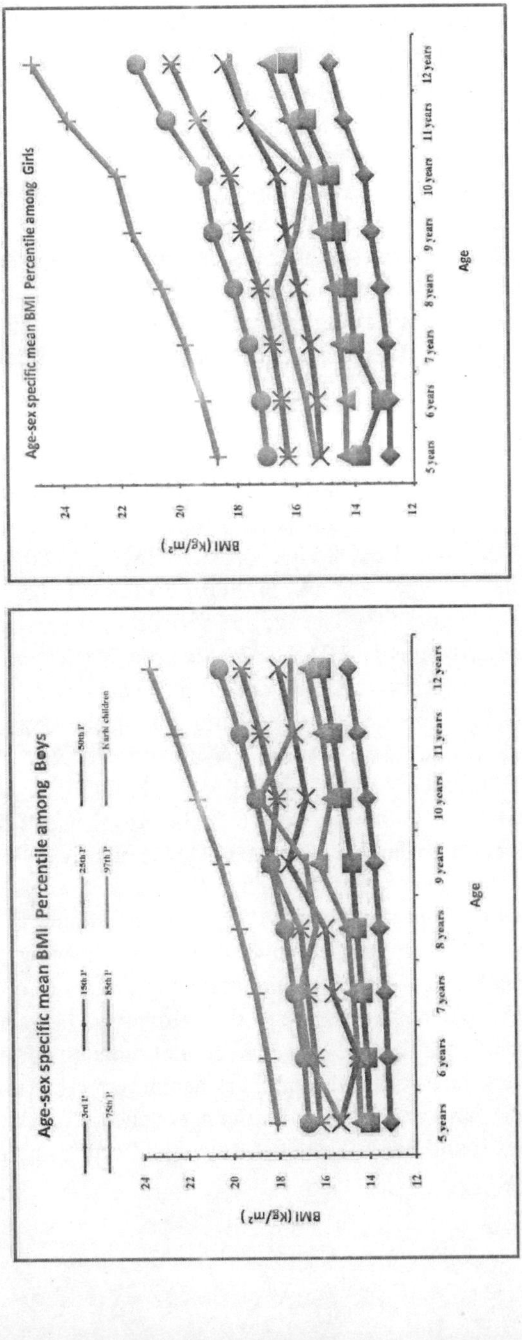

Figure 6.1: Age-sex specific mean BMI (kg/m²) comparison with the different percentiles of WHO (2007) reference population.

The overall prevalence of mild (14.52 per cent vs. 11.72 per cent) and severe (2.07 per cent vs. 0.83 per cent) thinness was found to be higher among girls than boys, but moderate thinness (4.60 per cent vs. 4.14 per cent) was found to higher among boys than girls ($p>0.05$). Age and sex-specific prevalence of mild thinness (Grade-I) was found to be higher among the boys and girls aged 12 years (18.52 per cent) and 7 years (25.81 per cent), respectively. The age- and sex-specific different grade of thinness prevalence was not showing any trends among the children. The sex difference in three different thinness grades was found to statistically insignificant using chi-square analysis ($p>0.05$). Age-sex specific prevalence of combined overall thinness (Grades I, II and III) prevalence among the Karbi children is also depicted in Figure 6. 2.

Prevalence of overweight and obesity

The age-sex specific prevalence of combined overweight-obesity among the tribal Karbi children is shown in Table 6.2. The prevalence of overall combined overweight-obesity (BMI $\geq$ 25.00 kg/m^2) was found to be 17.15 per cent and 11.20 per cent among boys and girls, respectively (Figure 6.3). The age and sex-specific overall prevalence of overweight-obesity (BMI $\geq$ 25.00 kg/m^2) was ranged from 6.25 per cent (5 years) to 33.33 per cent (in 7 years) and 2.63 per cent (10 years) to 22.58 per cent (8 years) among the boys and girls, respectively (Figure 6.3). Age-sex specific prevalence of overweight (BMI $\geq$ 25.00 kg/m^2) and obesity (BMI $\geq$ 30.00 kg/m^2) among the Karbi children is depicted in Table 6.2. The prevalence of overweight (14.17 per cent) is found to be greater than the obesity (3.33 per cent). The overall prevalence of overweight (17.15 per cent vs. 11.20 per cent) and obesity (3.76 per cent vs. 2.90 per cent) was found to be greater among the boys than the girls ($p>0.05$). The age-specific prevalence of overweight was found to be higher among 7 years (33.33 per cent boys; 22.58 per cent girls) ($p>0.05$). However, the age-specific prevalence of obesity and overweight are not showing any age related trends among the boys and the girls. The age and sex specific difference between the boys and the girls in the prevalence of overweight and obesity was found not to be statistically significant using chi-square analysis ($p>0.05$).

Table 6.2: Prevalence of thinness and overweight and obesity among the Karbi children of Karbi-Anglong, Assam

Age	*Number of subject*		*Thinness grade I (Mild)*		*Thinness grade II (Moderate)*		*Thinness grade III (Severe)*		*Overweight*		*Obesity*	
	Boys	*Girls*	*Boys*	*Girls*	*Boys*	*Girls*	*Boys*	*Girls*	*Boys*	*Girls*	*Boys*	*Girls*
5 years	32	27	5 (15.63)	7 (25.93)	3 (9.37)	0 (0.00)	0 (0.00)	1 (3.70)	2 (6.25)	3 (11.11)	0 (00.00)	1 (3.70)
6 years	32	25	3 (9.38)	2 (8.00)	3 (9.37)	3 (12.00)	1 (3.13)	0 (00.00)	5 (15.62)	2 (08.00)	6 (18.75)	1 (4.00)
7 years	27	31	2 (7.41)	8 (25.81)	0 (0.00)	0 (0.00)	0 (0.00)	1 (3.22)	9 (33.33)	7 (22.58)	1 (3.70)	2 (6.45)
8 years	36	27	3 (8.33)	4 (15.15)	2 (5.55)	1 (3.70)	0 (0.00)	0 (00.00)	4 (11.11)	6 (22.22)	0 (0.00)	1 (3.70)
9 years	27	33	2 (7.41)	5 (14.81)	0 (0.00)	2 (6.06)	1 (3.70)	0 (00.00)	6 (22.22)	2 (6.06)	0 (0.00)	0 (00.00)
10 years	25	38	4 (16)	5 (13.16)	0 (0.00)	3 (7.89)	0 (0.00)	1 (2.63)	8 (32.00)	1 (2.63)	2 (8.00)	0 (00.00)
11 years	33	24	4 (12.12)	2 (8.33)	2 (6.06)	1 (4.16)	0 (0.00)	1 (4.16)	5 (15.15)	3 (12.05)	0 (0.00)	1 (4.16)
12 years	27	36	5 (18.52)	2 (5.56)	1 (3.70)	0 (0.00)	0 (0.00)	1 (2.77)	2 (7.40)	4 (11.11)	0 (0.00)	1 (2.77)
Total	239	241	28 (11.72)	35 (14.52)	11 (4.60)	10 (4.14)	2 (0.83)	5 (2.07)	41 (17.15)	27 (11.20)	9 (3.76)	7 (2.90)

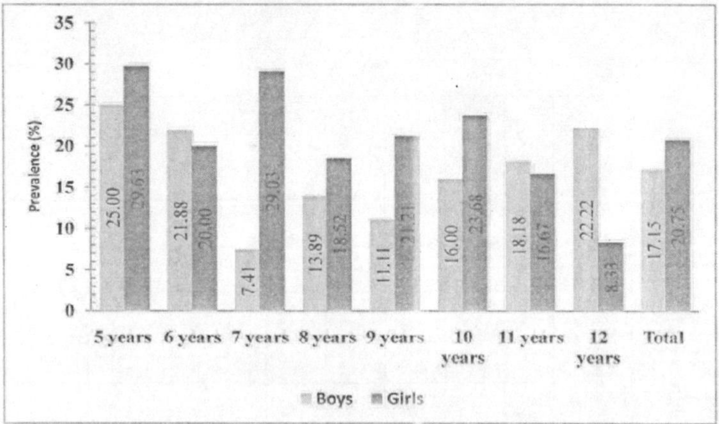

Figure 6.2: Age and sex specific prevalence of overall thinness (Grade I, II and III) among the Karbi children.

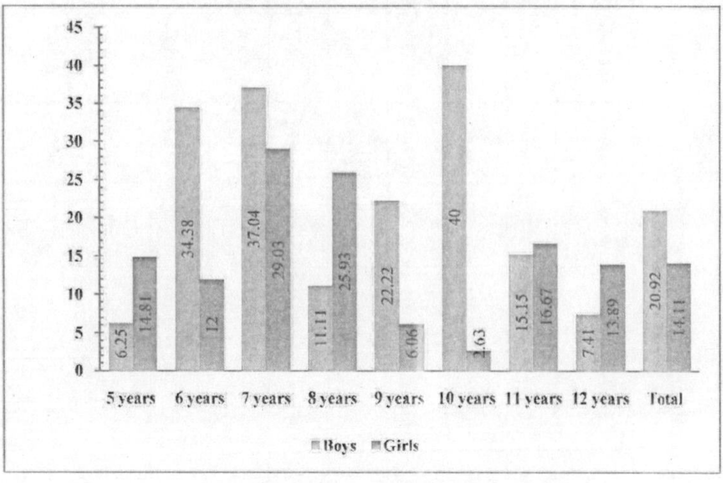

Figure 6.3: Age and sex specific prevalence of overweight (BMI $\geq$25.00 kg/m^2) among the Karbi children.

Discussion

Assessments of the undernutrition among children bear great significance in the developing countries such as India, where the vast majority of the populations are undernourished and underprivileged (Nandy *et al.*, 2005; Khor, 2008; Sen and Mondal, 2012; Varadharajan et al., 2013). Such nutritional assessment studies are found to be very important for the improvement of their health and overall development of the population concern. Recent studies have highlighted that India is currently undergoing a rapid transition on the economic, demographic, epidemiologic, nutrition and sociological fronts (Kapil and Sachdev, 2012; Varadharajan *et al.*, 2013). There is evidence of a decline in the prevalence of under-nutrition with a simultaneous escalation in overweight-obesity and associated non-communicable diseases in the population (Kapil and Sachdev, 2012; Varadharajan *et al.*, 2013). Despite of socio-economic development in the developing countries, the prevalence of under-nutrition is still remains an important public health issue that continues to be a major cause of morbidities and premature mortalities among children in India (Nandy *et al.*, 2005; Khor, 2008; Kapil and Sachdev, 2012; Varadharajan *et al.*, 2013). The use of anthropometry is the single most non-invasive and inexpensive technique plays very important role in the nutritional assessment and intervention programme in the population. The current concern and national policy response to tackling malnutrition in India is still primarily restricted to under-nutrition diagnosed and most of the large scale target intervention programmes are mainly concentrated on the anthropometry (Kapil and Sachdev, 2012). The conventional anthropometric measurement of stunting (low-height-for-age), underweight (low-weight-for-age) and wasting (low- weight-for-height) are commonly used to assess the prevalence of nutritional status and these indices reflect distinct biological processes among children (WHO, 1995, 2007; Nandy *et al.*, 2005; Kapil and Sachdev, 2012). However, these measures are generally overlapping in nature and unable to reflect the actual magnitude of under-nutrition in the population (Nandy *et al.*, 2005; Sen and Mondal, 2012). The BMI as measured by weight in kilogram (kg) divided by height in meter square (m^2) is an anthropometric measure that has been extensively used to determine the nutritional status (e.g., chronic energy deficiency and/or overweight-obesity) (WHO, 1995, 2007; Cole *et al.*, 2000, 2007). The use of BMI in relation to age (BMI-for-age) (e.g., thinness and overweight-obesity) has been recommended as the best surrogate anthropometric measure of thinness and overweight during adolescents (Cole *et al.*, 2000; 2007, WHO, 2007).

In the present study, an assessment of thinness, overweight and obesity among the Karbi children (aged 5-12 years) of Karbi-Anglong, Northeast India were undertaken using newly proposed cut-offs of Cole *et al.*, (2000, 2007). The use of newly proposed cut-offs is suggested to be encourage direct comparison of the prevalence of overweight-obesity and thinness that provide a new classification for the worldwide evaluation of public health purposes. Prior to this report, there were no such suitable thinness cut-off points were available for this age group (e.g., 2 to 18 years) except the National Centre of Health Statistics (NCHS) (Must *et al.*, 1991), National Health and Nutritional Examination Survey (NHANES) (WHO, 1995) and Centres of Disease Control and Prevention (CDC, 2000) (Kuczmarski *et al.*, 2002), Population for children and adolescents (WHO, 1995). It must be mentioned here that these cut-off points were derived from a multi-centre data from the United States, Great Britain, Hong Kong and the Netherlands including data from Brazil (developing country). Several studies have already reported the magnitude of the thinness among children and adolescents in Indian population (Chakraborty and Bose, 2009; Bisai and Manna, 2010; Mondal and Sen, 2010; Bisai *et al.*, 2010; Das and Bose, 2011; Singh and Mondal, 2013; Mondal, 2014; Sharma and Mondal, 2014; Thakur and Gautam, 2015; Tigga *et al.*, 2015) utilising these newly proposed cut-off points (Cole *et al.*, 2007). Hence, the proposed cut-offs point is valid for the assessment of thinness among Indian children and adolescents, including these Karbi children of Karbi-Anglong, India. It is generally agreed fact that different ethnic groups showed a significant variation in respect to the nutritional status in India. It was well established that in India, children belonging to the economically lower tribal and non-tribal, ethnic population were found to be poorer in nutritional status as compared to the general caste children (Som *et al.*, 2006; Mondal and Sen, 2010; Sen and Mondal, 2012). It is also apparent from this study, that there is a high prevalence of overall thinness (20.75 per cent girls; 17.15 per cent boys) observed among the school going children ($p>0.05$) (Figure 6.2). It has also been observed that several studies have already reported that boys were more sufferer than girls in thinness prevalence among children (Chakraborty and Bose, 2009; Bisai and Manna, 2010b;

Mondal and Sen, 2010). The comparison of overall thinness was found to be relatively lower than the reported studies among different tribal and non-tribal children in India. It is also evident that the prevalence of overall thinness was found to be higher among Bengalee (62.26 per cent; Chakraborty and Bose, 2009), rural children of Darjeeling (63.40 per cent; Mondal and Sen, 2010c), Kora-Modi tribe (67.23 per cent; Bisai *et al.*, 2010), Santal tribe (56.40 per cent; Das and Bose, 2011) and found lower among the Nepali children (45.15 per cent; Das and Banik, 2011) than the studied rural adolescent (49.10 per cent) in Darjeeling district using newly proposed cut-off (Cole *et al.*, 2007).

The recent trend shows that several developing countries are undergoing the rapid socio-economic and demographic transitions which leads to the major cause of acceleration of several preventable, non-communicable diseases, greater incidence of overweight-obesity and slightly lower prevalence of under-nutrition and thus existence the 'double burden' of malnutrition (Popkin, 2006; Varadharajan *et al.*, 2013). Several socio-economic, demographic and sedentary lifestyle changes are the crucial to ascertain the impact of overweight and obesity, because these problems are emerging at a time when under-nutrition remains a significant public health issue across the Indian populations (Khor, 2008; Wang *et al.*, 2009; Kapil and Sachdev, 2012; Varadharajan *et al.*, 2013). The result shows the prevalence of combined overweight obesity (BMI $\geq$ 25.00 kg/m^2) was found to be 20.92 per cent (boys) and 14.11 per cent (girls) (Table 6.2). Recent studies have reported the prevalence of overweight has increased in Indian children and adolescent of Ludhiana, Punjab (25.35 per cent) (Chhatwal *et al.*, 2004) (Delhi (22.43 per cent) (Sharma *et al.*, 2007), urban school children (16.81 per cent) (Ramachandran *et al.*, 2002), Punjabi children (14.31 per cent) (Sidhu *et al.*, 2006), affluent children-adolescent, Punjab (10.94 per cent) (Sidhu *et al.*, 2005), Pune (19.90 per cent) (Khadilkar and Khadilkar, 2004), Ludhiana, Punjab (12.70 per cent) (Aggarwal *et al.*, 2008) and Assam (8.13 per cent) (Sharma and Mondal, 2014) (Table 6.3).

Table 6.3: Prevalence of overweight among Indian school children

Population/ Group	*Study Area/region*	*Sample*	*Prevalence*	*Chi-value*	*Reference*
School children	Punjab	2008	25.35	8.42**	Chhatwal *et al.*, 2004
School children	Pune	1128	19.90	0.83	Khadilkar and Khadilkar, 2004
Affluent children	Delhi	4000	22.43	4.02*	Sharma *et al.*, 2007
Rural children	Assam	480	17.50	-	Present study
Urban children	India	4700	16.81	0.11	Ramachandran *et al.*, 2002
School children	West Bengal	753	15.1	0.87	Ghosh, 2011
Affluent children	Amritsar, Punjab	640	14.3	1.47	Sidhi *et al.*, 2005
Affluent children	Punjab	1000	10.94	9.40**	Sidhi *et al.*, 2006
Affluent children	Aligarh, Uttar Pradesh	660	13.2	2.99	Nawab *et al.*, 2014
School children	Assam India	1228	8.13	24.25**	Sharma and Mondal, 2014
School children	Ludhiana, Punjab	1000	12.7	4.52*	Aggarwal *et al.*, 2008

*$p<0.05$; **$p<0.01$

A comparison of the overweight-obesity prevalence of the different Indian children between Karbi children is also depicted in Figure 6.4. The results further suggested that the prevalence of overweight and obesity were found to be higher among boys than the girls ($p>0.05$). Several studies have already reported the similar trends that boys were found to be more sufferers in overweight-obese than their female counterparts (Chhatwal *et al.*, 2004). Therefore, the greater prevalence of overweight and obesity are generally apparent in the higher socio-economic strata due to change in dietary habit, sedentary lifestyle, less physical work and desk-bound activity. Several researchers have already suggested that the recent acceleration in the incidences of overweight-obesity in the population could be attributed as an environmental factors contributing to rapid urbanisation and changes in dietary patterns and lifestyles in Indian populations (Sharma *et al.*, 2007, Wang *et al.*, 2009; Kiranmala *et al.*, 2013; Varadharajan *et al.*, 2013). Therefore, the prevalence suggested that the magnitude is increasing not only the urban and sub-urban populations, but also in the rural and tribal populations in India. Therefore, the prevalence of childhood obesity and overweight could be a challenge for the governments and healthcare providers and would be certainly contribute to positive health gains and related co-morbidity in the foreseeable future.

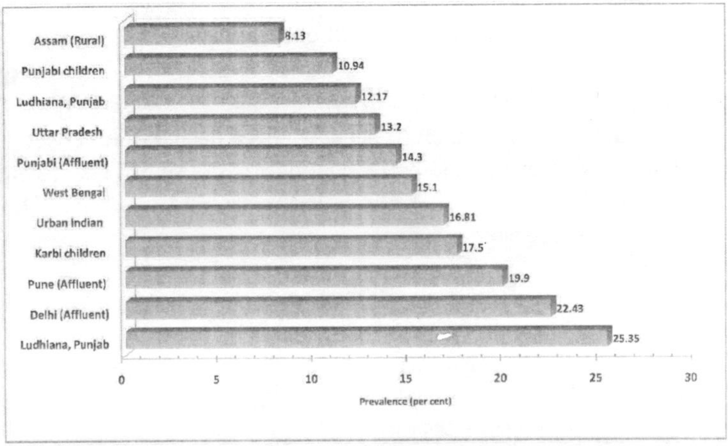

Figure 6.4: Prevalence of Overweigh among Indian children (Data Source Table 6.3).

Conclusion

The prevalence of greater degree of malnourishment is transversely persistent among the India children. The results further suggested that the prevalence of under-nutrition was assessed using the conventional anthropometric measure showed a major public health problem among the Karbi tribal school going children of Karbi-Anglong Assam, Northeast India. The comparison showed that the differences in the prevalence of under-nutrition in the population, suggesting that their genetic and ethnic makeup could play significant roles in such variation in prevalence. There were sex specific differences in the prevalence of under-nutrition between boys and girls, but girls were observed to be more affected by under-nutrition than boys in most cases. Since the ultimate goal of such nutritional studies is to improve human health, nutritional intervention becomes a priority for these children. Hence, the appropriate nutritional intervention programmes are required to improve the nutritional status of the Karbi tribal children. Dissemination of nutrition-related awareness among parents and community level while relatively greater improvement in weight than height among older children warns the future possibility of childhood adiposity even among the target populations. The intervention should be introduced in terms of healthy balance diet and protective nutrient to reduce the mortality, morbidity and ameliorate the nutritional status.

Acknowledgement

The authors are gratefully acknowledged the help and cooperation of school authorities and participant during the study. The extended help of Department of Anthropology, Assam University (Diphu Campus) is also acknowledged.

References

Ahmed, T, Hossain M, Sanin KI. 2012. Global burden of maternal and child under-nutrition and micronutrient deficiencies. *Ann Nutr Metab* 61: 8-17.

Bamji, MS, Murthy PVVS, Williams L, Vardhana Rao MVV. 2008. Maternal nutritional status and practices and prenatal, neonatal mortality in rural Andhra Pradesh, India. *Indian J Med Res* 127:44-51.

Bisai, S, Manna I. 2010. Prevalence of Thinness among Urban Poor Pre-School Children in West Bengal, India. *Sudanese J Pub Health* 4:193-98.

Bordoloi, BN, Sharma Thakur GC, Saikia GC. 1987. Tribes of Assam (Part-I). Tribal Research Institute, Guwahati.

Bose, K, Biswas S, Bisai S, Ganguli S, Khatun A, Mukhopadhyay A, Bhadra M. 2007. Stunting, underweight and wasting among Integrated Child Development Services (ICDS) scheme children aged 3–5 years of Chapra, Nadia District, West Bengal, India. *Matern Child Nutr* 3:216–221.

Chakraborty, R, Bose K. 2009. Very high prevalence of thinness using new international body mass index cut off points among 5-10 year old school children of Nandigram, West Bengal, India. *J Res Med Sci* 14:129–33.

Chhatwal, J, Verma M, Riar SK. 2004. Obesity among pre-adolescent and adolescents of a developing country (India). *Asia Pac J Clin Nutr* 13:231-5.

Chowdhury, SD, Chakraborty T, Ghosh T. 2008. Prevalence of undernutrition in Santal children of Puruliya district, West Bengal. *Indian Pediatr* 45:43–46.

Cole, TJ, Bellizzi MC, Flegal KM, Dietz WH. 2000. Establishing a standard definition for child overweight and obesity worldwide: international survey. *BMJ* 320 (7244):1240-43.

Cole, TJ, Flega KM , Nicholls D, Jackson AA .2007. Body mass index cut offs to define thinness in children and adolescents: international survey. *British Medical J* 335:- 194-201.

Das, S, Banik SD. 2011. Prevalence of thinness among Nepali speaking preschool children of Darjeeling using Body Mass Index cut-off points. *Italian J Pub Health* 8:241-46.

Das, S, Bose K. 2011. Prevalence of Thinness among Santal Preschool Children Using New Body Mass Index Cut-off Points. *J Anthropol*: Article ID 602782.

Ghosh, A. 2011. Rural-urban comparison in prevalence of overweight and obesity among children and adolescents of Asian Indian origin. *Asia Pac J Public Health* 23:928-35.

Hall, JG, Allanson JE, Gripp KW, Slavotinek AM. 2007. Handbook of Physical Measurements.New York: Oxford University Press.

Harris, EF, Smith RN. 2009. Accounting for measurement error: a critical but often overlooked process. *Arch Oral Biol* 54:S107-17.

Hien NN, Kam S. 2008. Nutritional status and the characteristics related to malnutrition in children under five years of age in Nghean, Vietnam. *J Prev Med Public Health* 41:232–40.

International Institute for Population Sciences (IIPS) and Macro International. 2007. National Family Health Survey (NFHS-3), 2005–06: India: Volume I. Mumbai: IIPS.

Kapil, U, Sachdev HP. 2012. Urgent need to orient public health response to rapid nutrition transition. *Indian J Community Med* 37:207-10.

Khadilkar VV, Khadilkar AV. 2004. Prevalence of obesity in affluent school boys in Pune. *Indian Pediatr* 41:857-8.

Khor, GL. 2008. Food-based approaches to combat the double burden among the poor: challenges in the Asian context. *Asia Pac J Clin Nutr* 17:111–15.

Kiranmala N, Das MK, Arora NK. 2013. Determinants of childhood obesity: need for a trans-sectoral convergent approach. Indian J Pediatr 80:S38-47.

Kuczmarski, RJ, Ogden CL, Guo SS, *et al.*, 2000. CDC growth charts for the United States: Methods and development. National Center for Health Statistics. **Vital Health Stat** 11(246), 2002.

Mandal, GC, Bose K. 2009. Assessment of overall prevalence of under-nutrition using composite index of anthropometric failure (CIAF) among preschool children of West Bengal, India, *Iranian J Pediatrics* 19:237–43.

Martorell, R, Young MF. 2012. Patterns of stunting and wasting: potential explanatory factors. *Adv Nutr* 3:227-33.

Mitra, M, Kumar PV, Chakrabartti S, Bharati P. 2007. Nutritional status of Kamar tribal children in Chhattisgarh. *Indian J Pediatr* 74:381–4.

Mittal, PC, Srivastava S. 2006. Diet, nutritional status and food related traditions of Oraon tribes of New Mal (West Bengal), India. *Rural Remote Health* 6(1):385 Epub 10 April 2006.

Mondal, N, Sen J. 2010a. Prevalence of stunting and thinness among rural adolescents of Darjeeling district, West Bengal, *India. Italian J Pub Health* 7:54–61.

Mondal, N, Sen J. 2010b. Prevalence of undernutrition among children (5-12 years) belonging to three communities residing in a similar habitat in North Bengal, India. *Ann Hum Biol* 37:198-216.

Mondal, N, Sen J. 2010c. Thinness is a major underlying problem among Indian children. *J Trop Pediatr* 56:456-8.

Mondal, N. 2014. Thinness is major underlying problem among rural adolescents ofNortheast India. *Nep J Pediatr Soc* 34:39-47.

Must A, Dallal GE, Dietz WH. 1991. Reference data for obesity: 85th and 95th percentiles of body mass index (wt/ht 2) and triceps skinfold thickness. *American journal of clinical nutrition* 53:839-46.

Nandy, S, Irving M, Gordon D, Subramanian SV, Smith GD. 2005. Poverty, child undernutrition and morbidity: New evidence from India. *Bull World Health Organ* 83:210–16.

Nawab, T, Khan Z, Khan IM, Ansari MA. 2014. Influence of behavioral determinants on the prevalence of overweight and obesity among school going adolescents of Aligarh. *Indian J Public Health* 58:121-4.

Popkin, BM. 2006. Global nutrition dynamics: the world is shifting rapidly toward a diet linked with noncommunicable diseases. *Am J Clin Nutr* 84:289–98.

Prasad, HK, Ekbote V, Rustagi V, Borade A, Chiplonkar S, Khadilkar V, Khadilkar AV. 2012. Performance of WHO growth standards on Indian children with growth related disorders. *Indian J Pediatr* 79:884-90.

Rahman, A, Chowdhury S. 2007. Determinants of chronic malnutrition among preschool children in Bangladesh. *Biosoc Sci* 39:161–73.

Ramachandran, A, Snehalatha C, Vinitha R, Thayyil M, Kumar CK, Sheeba L, Joseph S, Vijay V. 2002. Prevalence of overweight in urban Indian adolescent school children. *Diabetes Res Clin Pract* 57:185-90.

Ramachandran, P, Gopalan HS. 2011. Assessment of nutritional status in Indian preschool children using WHO 2006 Growth Standards. *Indian J Med Res* 134:47-53.

Ramachandran, P. 2007. Poverty nutrition linkages. *Indian J Med Res* 126: 249–61.

Rao, S, Kanade AN, Joshi SB, Sarode JS. 2012. Secular trends in growth of preschool children from rural Maharashtra, India. *J Health Popul Nutr* 30:420-30.

Seetharaman, N, Chacko TV, Shankar SLR, Mathew AC. 2007. Measuring Malnutrition -The Role of Z Scores and the Composite Index of Anthropometric Failure (CIAF). *Indian J Community Med* 32:35-39.

Sen, J, Mondal N. 2012. Socio-economic and demographic factors affecting the Composite Index of Anthropometric Failure (CIAF). *Ann Hum Biol* 39:129-36.

Sen, P, Bharati S, Som S, Pal M, Bharati P. 2011. Growth and nutritional status of preschool children in India: a study of two recent time periods. *Food Nutr Bull* 32:84-93.

Seth, A, Sharma R. 2013. Childhood obesity. *Indian J Pediatr* 80:309-17.

Sharma, A, Sharma K, Mathur KP. 2007. Growth pattern and prevalence of obesity in affluent school children of Delhi. *Public Health Nutr* 10:485-91.

Sharma, J, Mondal N. 2014. Prevalence of double nutrition burden among adolescent girls of Assam, Northeast India. *J Nepal Paediatr Soc* 34:132-37.

Sidhu, S, Kaur N, Kaur R. 2006. Overweight and obesity in affluent school children of Punjab. *Ann Hum Biol* 33:255-9.

Sidhu, S, Marwah G, Prabhjot. 2005. Prevalence of overweight and obesity among the affluent adolescent school children of Amritsar, Punjab. *Coll Antropol* 29:53-5.

Singh, J, Mondal N. 2013. Assessment of nutritional status: A case of tribal children in Assam, Northeast India. *J Nepal Paediatr Soc* 33:1-7.

Som, S, Pal M, Bhattacharya B, Bharati S, Bharati P. 2006. Socioeconomic differentials in nutritional status of children in the states of West Bengal and Assam, India. *J Biosoc Sci* 38:625-42.

Svedberg, P. 2011.How many people are malnourished? *Annu Rev Nutr* 31:263-83.

Thakur, R, Gautam RK. 2015. Nutritional status among boys and girls of a central Indian Town (Sagar). *Anthropological Review* 78: 197-212.

Touitou, Y, Portaluppi F, Smolensky MH, Rensing L. 2004. Ethical principles and standards for the conduct of human and animal biological rhythm research. *Chronobiol Int* 21:61-70.

Tigga, PL, Sen J, Mondal N. 2015. Association of some socio-economic and socio-demographic variables with wasting among pre-school children of North Bengal, India. *Ethiop J Health Sci* 25:63-72.

Ulijaszek, SJ, Kerr DA. 1999. Anthropometric measurement error and the assessment of nutritional status. *Br J Nutr* 82:165–77.

Varadharajan, KS, Thomas T, Kurpad AV. 2013. Poverty and the state of nutrition in India. *Asia Pac J Clin Nutr* 22:326-39.

Wang, Y, Chen HJ, Shaikh S, Mathur P. 2009. Is obesity becoming a public health problem in India? Examine the shift from under- to overnutrition problems over time. *Obes Rev* 10:456-74.

World Health Organisation (WHO). 2007. WHO child growth standards: head circumference-forage, arm circumference-forage, triceps skinfold-for-age and subscapular skinfold-for-age: methods and development.

World Health Organisation. 1995. Physical status: The use and interpretation of Anthropometry. Report of a WHO expert committee. WHO Geneva.

Zottarelli, LK, Sunil TS, Rajaram S. 2007. Influence of parental and socio-economic factors on stunting in children under 5 years in Egypt. East Mediterr Health J 13:1330–42.

7

Co-prevalence of Under-nutrition and Hypertension among Indian Tribal Populations

—Gautam K Kshatriya and Subhendu K. Acharya

Abstract

Nutritional transition has posed new challenges of triple burden of diseases. It is evident that obesity increases the risk of hypertension and therefore, the traditional perception about Indian tribal population which has been severely affected by under-nutrition is that they will remain free from such overweight related CVD health risks or such issues. In present context of knowledge, such traditional wisdom was observed to be false as under-nutrition stress also leads to hypertension. The present study finds very high prevalence of under-nutrition among all the six tribal population groups and mostly among the females. The mean BMI values at the lower border line of normal range for normal BMI among four of the six tribes further explains such observation. Along with, the significant gap between males and females with respect to mean weight in four age groups is evident of female marginalisation with respect to their access to basic nutritional needs. Along with, the findings show an alarming status of hypertension among both males and females which further explain an association of hypertension with increase in under-nutrition.

Introduction

Tribes and tribal societies in contemporary India are heading towards a paradigm shift with respect to their increasing exposure to the arena of development and related changes, leading towards a substantial transition in their dietary pattern, socio-cultural, economic and life style like aspects. But the tribal populations in various parts in India lacked the privilege from governmental as well as private bodies with respect to systematic observations of their various issues and further necessary actions. Such discordant conditions prevailed for a long time, even up to the end of 20th century. It has broadly been a traditional wisdom that the tribal people are contented within, easily avail all sorts of nutritious diet from forest, carry a good health from infancy to old age, total free from life style diseases, having their self-sufficient healthcare system and so on. In last few decades, though slowly, many facts evidently have surfaced contradicting these long standing biases and further revealing the alarming situation among these population groups. Particularly the reporting from health and healthcare dimensions indicated need of action in war steps. Further chronic poverty, nutrition- related health issues have come out in significant ways.

Studies related to health issues among the tribal populations of India have tended to focus on undernutrition (National Institute of Nutrition, ICMR, 1971, Basu *et al.*, 1990; Haque, 1990; Mohapatra, 1990; National Institute of Nutrition, ICMR, 2000; Census of India, 2001; Radhakrishna, 2004; National Institute of Nutrition, ICMR, 2004; Kar *et al.*, 2007). However, it has become necessary to conceptualise such studies which lay emphasis on the assessment of the health status of various tribal groups with respect to undernutrition and its related morbidity factors. For example, it is evident that obesity increases the risk of hypertension and therefore, the traditional perception about Indian tribal population which is severely affected by undernutrition is that they will remain free from such overweight related health issues, was proved false as undernutrition stress (Stini, 1985) also leads to hypertension.

Like all other developing countries, large scale urbanisation/ modernisation has been taking place in India with effective changes in the lifestyles leading to appreciable increase in the prevalence of chronic metabolic conditions like hypertension, cardiovascular diseases (CVD), diabetes, metabolic syndromes (Kusuma *et al.*, 2001; Kerketta, *et al.*, 2009; Sachdev, 2011; Manimudra, 2011; Madani, 2011;

Kapoor, *et al.*, 2012; Ramalingam, *et al.*, 2012; Meshram, 2012). But conventionally, it has been believed that tribal population groups are less exposed to socio-cultural and economic changes, so they are not unaffected by the process of socio-economic and ecological changes.

All the above views have been explained by findings reported in a number of studies conducted among different tribal groups in India. So, with the increasing status of non-communicable diseases among Indian tribes there is a growing concern about the increased metabolic risks among the tribal populations in the country.

It is important to note that the anthropometric measures among the tribal populations in India can be associated with metabolic risk factors, to predict their metabolic and cardiovascular health. BMI has been considered as an important physiological-index for mapping nutritional status. Associations of BMI with metabolic health risk factors like blood pressure (SBP and DBP) among Indian tribal populations indicate the association of undernutrition prevalence with cardio-vascular health (Kusuma *et al.*, 2001; Kerketta *et al.*, 2009; Sachdev, 2011; Manimudra, 2011; Madani, 2011; Kapoor *et al.*, 2012; Ramalingam *et al.*, 2012; Meshram, 2012). Again, it can further be said that many of the tribal populations are becoming susceptible to various life style risk factors. Therefore, it is worth investigating the anthropometric measures among the tribal population groups of India in order to predict metabolic health risk factors in a more meaningful way.

Material and Methods

The present study was undertaken to understand the pattern of height, weight, Body Mass Index (BMI) and prevalence status of undernutrition, hypertension like different health risk factors; among six tribes; the Santals, the Oraons and the Koras in Birbhum district of West Bengal and the Santals, the Bhumijs and the Bathudis in Mayurbhanj district of Odisha. Here, it is noteworthy that West Bengal and Odisha being two tribal dominated states, the population of different tribes of these two states constitutes 7.8 per cent and 25.7 per cent of the total populations respectively; additionally the tribal populations of West Bengal comprise of 38 tribal groups, while those in Odisha comprise of 62 tribal groups (Census 2011).

Area and People

Birbhum district

The scheduled tribe (ST) population in Birbhum district constitutes 6.74 per cent of the total population (Census, 2001).

Santal tribe in West Bengal speaks Santali language. They also speak Bengali, Oriya and Hindi. Santals of West Bengal have a script of their own named 'Ol chiki'. Traditionally the tribe is dependent on forest and they also practice agriculture. During festive occasions, Santal women dress in the red bordered white sari and dance in the line sequence. Santals believe in soul and ancestral spirit and pay respect to the ghosts and spirits like Kal Sing, Lakchera, Beudarang etc. Animal sacrifice is a common practice amongst this tribe. Karam festival which falls in the month of September and October is a well-known festival. The Santal people also observe rituals during hunting called Disumsendra.

It was observed that the occupation pattern of Santals is shifting from agriculture to migrant labour and from forest product collection to agriculture.

Koras are small tribal community in comparison to Santals in West Bengal. They speak the Indo-Aryan language Khotta among themselves. Traditionally, Koras depend on manual labour and agriculture for their livelihood. But in present findings, they are observed to be shifting towards daily wage labour. They are mostly non-vegetarian. Besides being cultivators and agricultural labourers, the Koras also get engaged as share croppers, construction workers and woodcutters. They have a traditional community council to control their social affairs. The religion of Koras is a mixture of their traditional faith and Hinduism.

Alcoholic behaviour was found prevalent among the Kora. Locally distilled liquor is predominantly consumed, but a shift towards foreign liquor can also be observed among the industrial and migrant labour class of this tribe.

Unlike Santals, Koras are yet to witness significant occupational shift from traditional patterns. Availability of facilities like road communication, piped drinking water etc. is not found in all the studied Kora villages. Access to electricity was found in all the Kora inhabiting villages.

Oraons is the major tribal group in West Bengal also known as Kurukh tribes, as they speak Kurukh language. They are found in the states of Jharkhand, Bihar, West Bengal and Odisha. Oraons are believed to have first settled in the Chhotanagpur Plateau and are considered to be the second largest tribes in Bihar and Jharkhand. Traditionally, they depend on agriculture for earning their livelihood. Some of them work as daily wage labourers and collect forest produces. Oraons are further divided into sub-groups like Kudas and Kisans. They follow patrilineal family customs. There are a total of 14 clans among Oraon.

The Oraons have adapted new occupational and livelihood patterns like industrial and migrant labour along with forest product collection and agriculture. Moderate to heavy consumption of both local as well as foreign liquor was witnessed among Oraon. Elderly and middle aged females were also fond of liquor.

Mayurbhanj District

Mayurbhanj district of Odisha is mainly a tribal-dominated district. Out of 62 types of tribes in Odisha, Mayurbhanj alone have 53 types. As per the Census 2001, the ST population in the district constitutes 56.6 per cent of the total population. In the present study, Santal, Bhumij and Bathudi tribes were studied from Mayurbhanj district in Odisha.

Santals in Odisha speak Mundari language. Santals live in scattered settlements. The Santal families are mostly nuclear and patrilineal. The sex ratio among Santals is 976 females per 1000 males. The total literacy rate is 40.46 per cent; whereas it is 55.86 per cent among males and 24.75 per cent among females.

Bhumijs of Odisha speak Bhumij language which is an Astro-Asiatic language. As per Census of India (2001) the sex ratio of the tribe is 984 females per 1000 males. The literacy rate among Bhumijs is 36.48 per cent, out of which males comprise 51.09 per cent and females comprise 21.72 per cent. Bhumijs live in scattered settlements. The Bhumij families are mostly nuclear and patrilinial.

Bathudi people speak Odia language which is an Indo-Aryan language. As per Census of India (2001), the sex ratio among the Bathudis is 1003 females per 1000 males. The total literacy rate among Bathudis is 49.56 per cent; whereas male literacy is 66.25 per cent and females literacy is 33.03 per cent. Bathudis live in uniclan and homogenous settlements. Bathudi families are mostly nuclear, patrilocal and patrilineal.

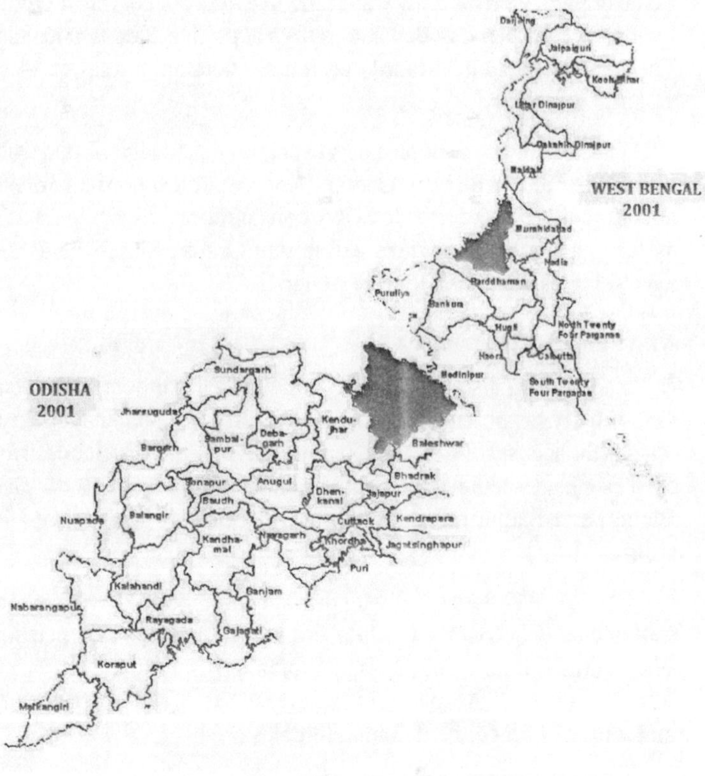

Map 7.1: Geographic location of selected districts in West Bengal and Odisha states on India Map.

Distillation and consumption of liquor at local level named as 'handia', 'mahuli', 'chauli' etc. was observed among all the three tribes. People preparing liquor from different fruits like mango, banana, jackfruit etc was also witnessed. Many tribal families earn their livelihood by selling home distilled liquor. The habit of consuming locally distilled liquor has largely been replaced by foreign made types among the people working in factories or as migrant and daily wage labourer in towns and cities.

Study Design

Selection of Sample

Villages in each of the two states which had a preponderance of the chosen tribes were selected for the study, while also taking into consideration the fact that the selected tribal villages exhibited differential developmental activities. From these villages, adult male and female subjects were selected through random sampling. The formula provided by Danial was used to estimate the sample size (Danial, 1999). The prevalence rate for calculating the sample size was based on the prevalence of obesity in Indian population (Mishra and Khurana, 2009).

The study comprised of four broad age groups of 21-30 years, 31-40 years, 41-50 years and 51-60 years. Criteria used to exclude individuals from the sample were: growth and developmental disorders, severe health problems in last one year and existence of any secondary cause of hypertension. The sample size for the present study was tested at five per cent level of significance with a power of 80 per cent.

The study was conducted between June 2011 and July 2013 in different phases. All the subjects were picked up randomly and were not related to each other.

Informed written consent from the subjects participating in the study was obtained prior to the actual commencement of the study. Prior ethical clearance from the Department Ethical Committee was obtained to conduct the research.

Anthropometric and Metabolic Variables

The primary information about the subjects who participated in the study, like name of the tribe, age and sex was recorded in a structured schedule. Standard techniques were followed while taking all the anthropometric measurements (Lohman, 1988). Standing height and weight was measured to the nearest 0.1 cm and 0.1 kg respectively. Stature was measured by using movable anthropometer and weight by using an Omeron Karada Scan Body Composition Monitor. The participants were encouraged to remove their shoes and heavy clothing before giving the measurements. BMI was calculated as weight in kilogram (kg) divided by height in metre squared (m^2) - kg/m^2. BP's, both systolic and diastolic were recorded by using a standard mercury sphygmomanometer on the right arm of the subjects after ensuring a minimum five minutes rest before the measurement. Two measurements were taken with a time gap of at least five minutes and the average of the two was recorded. Two trained anthropologists were involved in the data collection process. In order to avoid measurement and data entry bias, all the measurements were taken by one anthropologist while all the data was entered in the data sheet by the other. During the study the same instruments were used for taking the measurements of all the participants. Participants with Systolic Blood Pressure (SBP) $\geq$140 mmHg and Diastolic Blood Pressure (DBP) $\geq$90 mmHg were considered to be hypertensive as per the International Diabetic Federation (IDF) criteria (Zimmet and Alberti, 2006). Subjects with blood pressure $\geq$140/90 mmHg were considered as hypertensive. An individual condition with a BMI of <18.5 kg/m^2 was considered as undernutrition per the World Health Organisation (WHO) guidelines for Asian Populations (WHO, 2004).

Statistical Analysis

After incorporating and systematising the data in Microsoft Excel 2007, further analysis was carried out using SPSS Version 16.0 for Windows, SPSS Inc., Chicago, Illinois, USA. In order to maintain the integrity of the sample data, double entry of the data was done. The validity of the data was maintained by multiple-time cross-check so that the positive, negative and decimal characters of the numbers in the data could be ensured. Age group-wise and tribe-wise descriptive statistics such as mean and Standard Deviation (SD) for each variable was estimated for the selected anthropometric variables. Analysis was done to understand the prevalence percentage of selected risk factors such as high SBP, high DBP, High BP and undernutrition status in age groups and selected tribes. Mean values for normal and each risk category among males, females were calculated and were presented as error bar graphs at 95 per cent confidential level.

Results

Table 7.1 represents the descriptive statistics of the selected variables in four age groups. It is observable that there is an increasing trend in average height from older generation (51-60 yrs) to younger generation (21-30 yrs). But it is also observed that the average weight through generation does not show a pattern and prevails inconsistently. Further, the SBP and DBP show a trend of raising pattern along with increase in age. The mean SBP for 21-30 years age group was observed to be 119.86 (SD=14.34) mmHg which reaches to 136.05 (SD=22.36) mmHg for 51-60 years age group. Similarly, the mean DBP for 21-30 years age group was observed to be 75.49 (SD=11.10) mmHg which reaches to 83.88 (SD=13.85) mmHg for 51-60 years age group.

Considering BMI, it was observed that the mean BMI for all the four age groups lies nearer to lower cut off for normal BMI range (BMI cut off for normal status ≥18.5 to <23 kg/m^2)) which is indicative of high undernutrition condition among the population.

Table 7.1: Descriptive statistics of Height, Weight, Blood Pressure and BMI in four age groups

Variable	*Age group (years)*	*Sample size*	*Mean+SD*
Height (in cm)	21-30	372	155.03±8.98
	31-40	358	154.88±8.69
	41-50	341	153.80±7.68
	51-60	363	153.25±8.91
Weight (in kg)	21-30	372	46.85±9.83
	31-40	358	47.04±9.66
	41-50	341	46.08±9.27
	51-60	363	45.18±9.81

(Contd...)

SBP (mmHg)	21-30	372	119.86±14.34
	31-40	358	122.98±16.41
	41-50	341	129.49±23.65
	51-60	363	136.05±22.36
DBP (mmHg)	21-30	372	75.49±11.10
	31-40	358	79.20±11.47
	41-50	341	82.36±13.75
	51-60	363	83.88±13.85
BMI	21-30	372	19.38±2.95
	31-40	358	19.46±2.91
	41-50	341	19.65±6.26
	51-60	363	19.07±3.16

Tribe-wise descriptive statistics for height, weight, blood pressure and BMI is presented in Table 7.2. Here, 'Santal (O)' is abbreviation for Santal tribe of Odisha and 'Santal (WB)' is for Santal tribe from West Bengal.

Table 7.2: Descriptive statistics of Height, Weight, Blood Pressure and BMI in six tribal groups

Variables	*Tribal group*	*Mean+SD*
Height (in cm)		
	Bathudi	151.6+7.86
	Bhumij	155.25+7.92
	Kora	152.78+8.56
	Oraon	154.91+9.07
	Santal (O)*	155.74+8.67
	Santal (WB)*	155.17+8.83
Weight (in kg)		
	Bathudi	42.69+9.17
	Bhumij	49.34+10.02
	Kora	42.97+8.19
	Oraon	45.2+9.52
	Santal (O)*	49.63+9.42
	Santal (WB)*	47.84+9.17

(Contd...)

SBP (mmHg)		
	Bathudi	129.19+21.06
	Bhumij	129.33+23.44
	Kora	124.71+21.64
	Oraon	127.67+18.25
	Santal (O)*	125.61+15.98
	Santal (WB)*	125.74+21.31
DBP (mmHg)		
	Bathudi	81.11+14.22
	Bhumij	79.68+13.61
	Kora	79.78+12.88
	Oraon	82.28+12.46
	Santal (O)*	79.12+12.29
	Santal (WB)*	79.13+12.17
BMI		
	Bathudi	18.57+3.29
	Bhumij	20.31+3.05
	Kora	18.27+2.55
	Oraon	19.22+7.13
	Santal (O)*	20.25+2.92
	Santal (WB)*	19.68+2.91

* Santal (O) is Santal tribe from Odisha and Santal (WB) is Santal tribe from West Bengal.

It is apparent that mean height for Koras (151.6 ±7.86 cm) and Bathudis (152.78±8.56 cm) is distinctly less than remaining four tribes (Bhumij= 155.25±7.92 cm), Santal (O) = 155.74±8.67 cm, Santal (WB) = 155.17±8.83 cm, Oraon= 154.91±9.07 cm which show a close similarity in mean height. The mean weight among Koras and Bathudis stands distinctly low from rest of the four tribes. The mean SBP for all the six tribes is comparatively high than average Indian SBP (120-123 mmHg) with Bathudis, Bhumij and Oraons at the elevated level. Similarly, the mean DBP among Bathudis and Oraons stands at elevated level than other four tribes.

Important observations came out with respect to BMI. It is evident that the mean BMI for Koras (18.27±2.55 kg/m^2) is lower than cut off mark of the normal BMI range. Bathudis are at the border line (18.57±3.29). The average BMI for two other West Bengal tribes are observed to be closer to lower cut off of normal BMI range.

Figure 7.1 (A and B) exhibit the prevalence of various risk conditions of selected variables through age groups and tribes. The prevalence of hypertensive BP shows an increasing trend with increase in age. Further, the prevalence of undernutrition (BMI<18.5 kg/m^2) decreases from older age groups to younger age groups, the prevalence rate was observed to be very high in all the four groups varying from 37.37 per cent among 21-30 age group to 44.90 per cent among 51-60 age group, with 41.42 per cent overall prevalence.

With respect to tribe-wise prevalence percentage of various risk conditions of selected variables, it shows an alarming condition. The prevalence percentage of hypertensive BP was observed among 12.55 per cent of selected samples of the six tribes with 9.75 per cent among Oraons to 14.58 per cent among Santal of Odisha.

It is evident that more than 50 per cent of studied tribes namely, Bhumij, Kora and Santals of West Bengal have undernutrition (BMI<18.5 kg/m^2) with an average 41.42 per cent overall prevalence.

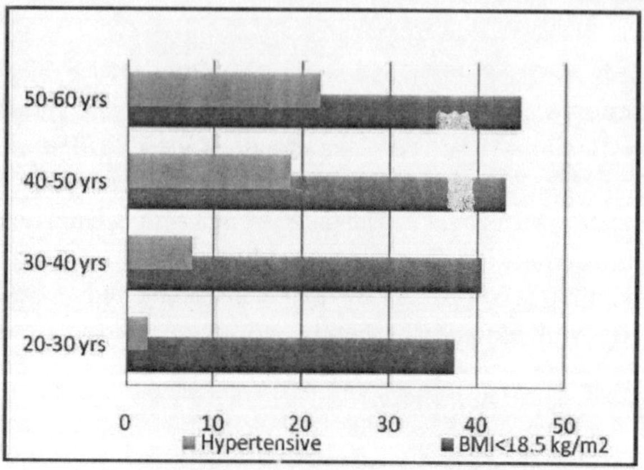

Figure 7.1A: Prevalence of undernutrition and hypertension among four age groups

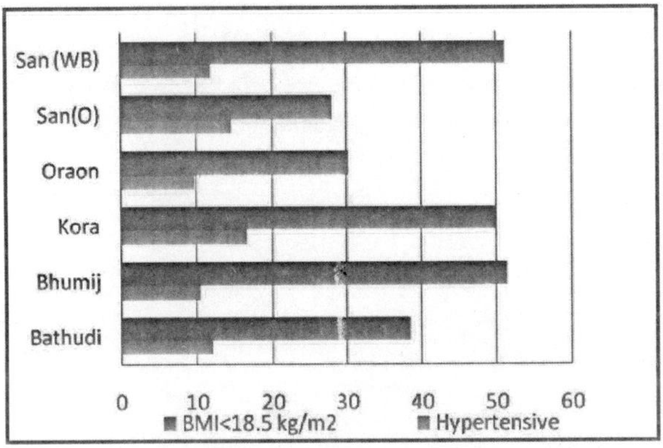

Figure 7.1B: Prevalence of undernutrition and hypertension among six tribes

Figure 7.2 and figure 7.3 represent the age group-wise patterns of variation of weight and height among males and females. It is clear that males are significantly taller and heavier as compared to females. Further, it should be noted that the younger males are taller and heavier as compared to their aged counterpart; no such trend is visible among females.

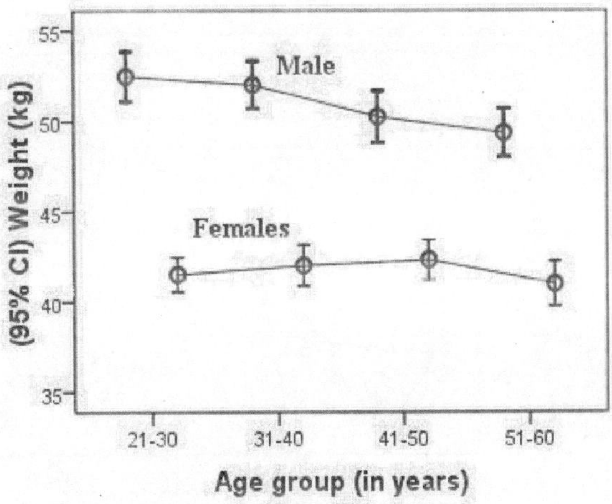

Figure 7.2: Age group-wise patterns of variation among males and females in weight.

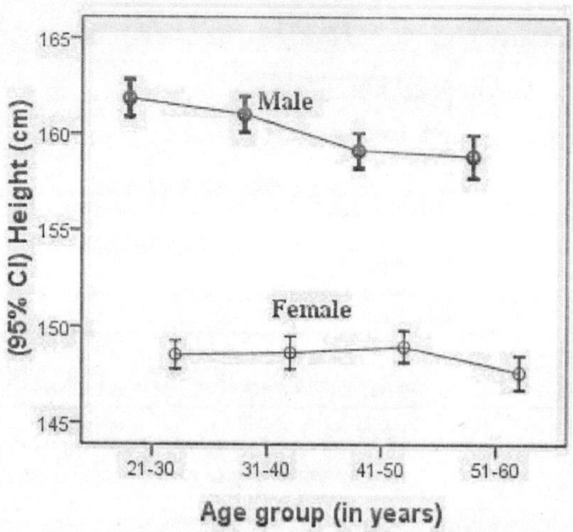

Figure 7.3: Age group-wise patterns of variation among males and females in height.

On the other hand, while analysing the pattern of hypertension (Figure 7.4 and Figure 7.5) in the six tribal groups, it was found that the prevalence of hypertensive SBP and DBP is higher among females than males especially in aged groups.

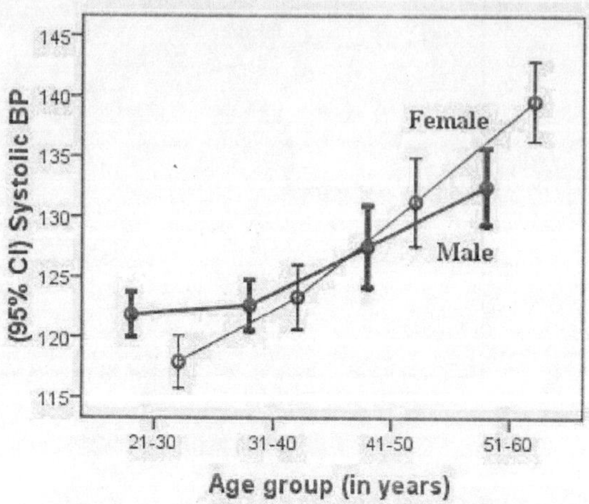

Figure 7.4. Age group-wise patterns of variation among males and females in hypertensive systolic BP.

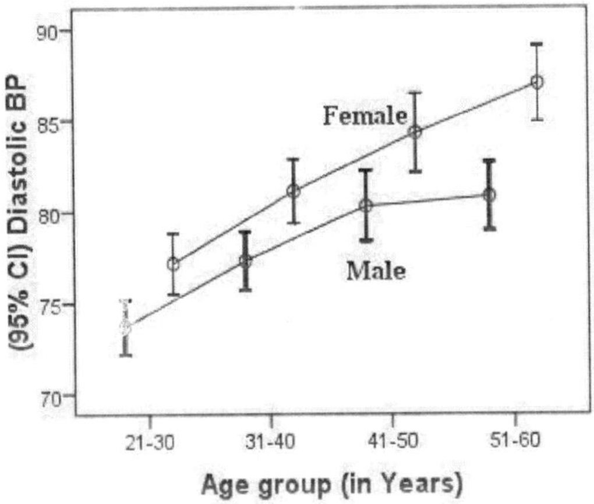

Figure 7.5. Age group-wise patterns of variation among males and females in hypertensive diastolic BP.

The BMI trend (Figure 7.6) among males and females in four age groups also exhibits a comparatively poor nutritional status among females than males.

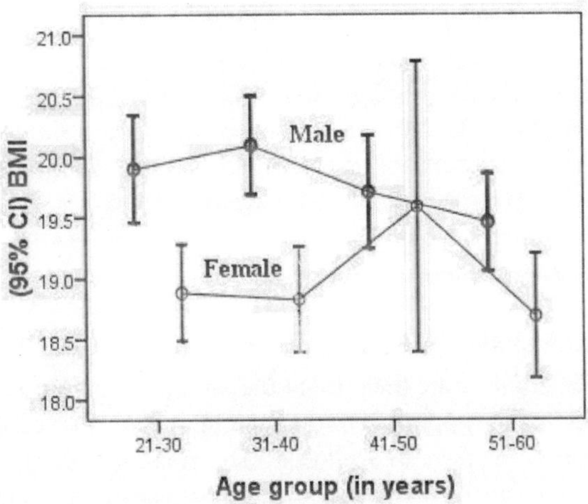

Figure 7.6: BMI trend among males and females in four age groups.

Conclusion

Scientific research studies, government reports and various census data have highlighted the high prevalence of undernutrition among Indian tribes (Adak *et al.*, 2006; Gautam *et al.*, 2006 and Gautam, 2008). Broadly these findings have explained the lack of access to proper food among Indian tribal population groups and their socio-economic marginalisation. The present study finds very high prevalence of undernutrition among all the six tribal population groups and mostly among the females. The mean BMI values at the lower border line of normal range of BMI among four of the six tribes further explains it. Similarly, the significant gap between males and females with respect to mean weight in four age groups is evident of female marginalisation with respect to their access to basic needs. Along with, the findings show an alarming status of hypertension both among males and females which further explain an increasing association of hypertension to increase in undernutrition. Such conditions put females in maximum vulnerable position. The micro level findings of trends in height and weight among 41-50 years and 51-60 years age groups in males and all the four age groups in females show continued and constant poor health. So an alarming figure comes out that of co-prevalence of undernutrition and hypertension. Researches with respect to prevalence of hypertension among Indian tribes are scanty. Further such finding showing co-prevalence of undernutrition and hypertension also provide insight about tribal health status and changing pattern. So it is need of the time to consider all such aspects at policy level which have long been neglected with the traditional wisdom that tribal population groups are not affected by life style diseases.

Acknowledgements

The Authors are thankful to Indian Council of Medical Research for financial support. The help rendered by Professor Satwanti Kapoor, University of Delhi and Dr Arnab Ghosh, Visva Bharati, Santi Niketan in formulation of the research problem is greatly appreciated. We express our gratitude to Miss Hemlata Dewangan and Mr Gajanan for data collection.

References

Adak, DK, Gautam RK, Bharati S, Gharami A, Pal M and Bharati P. 2006. Body mass index and chronic energy deficiency of adult males of central Indian populations. *Human biology*, 78(2): 161-178.

Andrews, FJ, Nolan JP. Critical care in the emergency department: monitoring the critically ill patient. *Emerg Med J* 2006; *23:* 561–64.

Babu, BV, Kusuma YS, Naidu JM. Distribution of blood pressure and influence of subcutaneous fat on systolic and diastolic levels in a tribal population. *J Ind Med Asso*1996; *94*:289-293.

Basu, SK, Jindal A, Kshatriya GK. Genetic and socio-cultural determinants of tribal health: A primitive Kutia Kondh tribal group of Phulbani district, Orissa. ICMR Final Report: NIHFW; 1990.

Bose, K, Bisai S, Chakraborty F. Age Variations in Anthropometric and Body Composition Characteristics and Underweight Among Male Bathudis – A Tribal Population of Keonjhar District, Orissa, India. *Coll. Antropol* 2006; 30: 771–75.

Chakraborty, R, Bose K. Anthropometric characteristics and nutritional status of adult Oraon men of Gumla District, Jharkhand, India. *The Internet Journal of Biological Anthropology* 2008; 2: 0–6.

Daniel, WW. Biostatistics: A foundation for Analysis in the Health Sciences. 7th ed. New Work: John Wiley & Sons; 1999.

Demographic status of schedule tribe population of India [data base on the internet]. New Delhi: Ministry of tribal affairs. Government of India [cited 2014 April 02] Available from: http://www.tribal.gov.in/WriteReadData/CMS/Documents/201306110208002203443 Demographic Status of Scheduled Tribe Population of India.pdf.

Gautam, RK, Adak DK, Gharami AK and Dutta T. 2006. Body Mass Index in Central India: Inter District Variation. *Anthrop. Anz*, 64 (4), 1-15

Gautam, RK. 2007b. Biosocial covariates of Adult male Body Mass Index in central India, *J. Biosoc. Sci. doi*, 10.

Gautam, RK. 2008. Traditional occupations and nutritionals adaptation among central Indian population. *J Biosoc Sci* 40(5):697-723.

Haque, M. Height, Weight and Nutrition among the six tribes of India. In: Chaudhuri B, editor. Cultural and Environmental Dimension on Health. New Delhi: Inter-India Publication; 1990. pp. 192-206.

Kapoor, AK, Saluja K, Verma D and Kapoor S. Predictors of Hypertension among Adult Tribal Males of India. *International Journal of Tropical Disease and Health* 2012; 2: 241-56.

Kar, GC, Sarangi L, Nanda A. A Study of Food Related Nutritional Deficiency in KBK Districts of Orissa. Planning Commission: Govt. of India; 2007.

Kerketta, AS, Bulliyya G, Babu BV, Mohapatra SS, and Nayak RN. Health status of the elderly population among four primitive tribes of Orissa India: a clinico-epidemiological study. *Z Gerontol Geriatr* 2009; *42*: 53-9.

Kusuma, YS, Babu BV, Naidu JM. Blood pressure levels among a few cross-cultural populations of Visakhapatnam district, Andhra Pradesh, India. *Annals of Human Biology* 2002; *29*: 502-512.

Kusuma, YS, Babu BV, Naidu JM. Group- and sex- specific effects of age, body composition and pulse rate on blood pressure variability in some cross-cultural populations of Visakhapatnam district, South India. *Journal of Cardiovascular Risk* 2001; *8*: 337-7.

Kusuma, YS, Das PK. Hypertension in Orissa, India. A cross-sectional study among some tribal, rural and urban populations. *Public Health* 2008; 22: 1120-23.

Lohman, TG, Roche AF, Martorell R. Anthropometric Standardisation Reference Manual. Human Kinetics Books: Chicago; 1988.

Mahapatra, D, Das J. Nutritional Ecosystems of Orissa tribals. In: Chaudhuri B, editor. Cultural and Environmental Dimension on Health. New Delhi: Inter-India Publication. 1990.

Mandani, B, Vaghani B, Gorasiya M, Patel P. Epidemiological Factors Associated With Hypertension Among Tribal Population in Gujarat. *National Journal of Community Medicine* 2011; 2: 133-5.

Manimunda, SP, Sugunan AP, Benegal V, Balakrishna N, Rao MV, Pesala KS. Association of hypertension with risk factors and hypertension related behaviour among the aboriginal Nicobarese tribe living in Car Nicobar Island, India. *The Indian Journal of Medical Research* 2011; 133: 287-93.

Meshram, II, Arlappa N, Balkrishna N, Rao KM, Laxmaiah A, Brahmam GNV. Prevalence of hypertension, its correlates and awareness among adult tribal population of Kerala state, India. *Journal of Post Graduate Medicine* 2012; 58: 255-61.

Misra, A, Khurana L. The Metabolic Syndrome in South Asians: Epidemiology, Determinants, and Prevention. *Metabolic Syndrome and Related Disorders* 2009; *7*: 497-514.

Mukhopadhyay, B, Mukhopadhyay S. Blood pressure and its biocultural correlates among the Lepchas of Sikkim, India: a micro level epidemiological study. *Collegium antropologicum* 2001; *25*: 97-110.

National Institute of Nutrition; ICMR, Hyderabad. Annual Report. ICMR; 1971.

National Institute of Nutrition; ICMR, Hyderabad. Annual Report. ICMR; 2004.

National Institute of Nutrition; ICMR. Special report on elderly and adolescents diet and nutritional status. ICMR; 2000.

Radhakrishna, R, Ravi C. Malnutrition in India: Trends and Determinants. *Economic and Political Weekly* 2004; *39*: 14-20.

Ramalingam, S, Murali A, Seethalakshmi A *et al.*, Cardiovascular Disease Risk Factors in a Tribal Population of Nilgiris. *Nat.J.Res.Com.Med* 2012; *1*: 90-5.

Sachdev, B. Prevalence of hypertension and associated risk factors among Nomad Tribe groups Screening of Hypertension, Adiposities and ABO Blood Group among Select Nomad Tribes of Rajasthan, India. *Antrocom Online Journal of Anthropology* 2011; *7*: 181-9.

Schall, JI. Sex Differences in the Response of Blood Pressure to Modernisation. *American Journal of Human Biology* 1995; *7*: 159-72.

Stini, WA. Growth rates and sexual dimorphism in evolutionary perspective. In: Gilbert RI, Mielke JH, editors. The Analysis of Prehistoric Diets. New York: Academic Press. 1985. pp. 191-226.

Swai, AB, Kitange HM, Masuki G, Kilima PM, George KM, Alberti, Mclarty DG. Is Diabetes Mellitus Related to Undernutrition in Rural Tanzania? *BMJ: British Medical Journal* 1992; *305*: 1057-62.

Van der Sande, MA, Ceesay SM, Milligan PJ, Nyan OA, Banya WA, Prentice A, McAdam KP, Walraven Gijs EL. Obesity and undernutrition and cardiovascular risk factors in rural and urban Gambian communities. *American Journal of Public Health* 2001; 91: 1641-4.

World Health Organisation. Appropriate body-mass index for Asian populations and its implications for policy and intervention strategies. WHO expert consultation. *The Lancet* 2004; *363*: 157-63.

Zimmet, PZ, Alberti KG. Introduction: Globalisation and the noncommunicable disease epidemic. *Obesity (Silver Spring)* 2006; 14:13.

8

Prevalence of Anemia among Tribals in Mayurbhanj District of Odisha

—*Manoranjan Mohapatra,*
K.C. Satapathy and P. K. Patra

Abstract

Nutritional anemia is a major public health problem among the Indian populations especially in the rural and tribal belts. Odisha has good health indicators in comparison to other states of India. However, the health of tribals of the state is poor. The primary objective of the study is to estimate the prevalence of anemia among tribals of different age groups of Mayurbhanj district. A cross-sectional survey was done among a sample of 122 tribals from different age groups in three clusters of Balidiha Gram Panchayat of Mayurbhanj District on a random basis in 2013. Anemia was diagnosed by estimating the hemoglobin concentration in the blood with the use of Sysmax KX-21 Haematolgy Analyser. The mean hemoglobin of the three study groups was 11.51±2.01 gm per cent whereas in case of male mean haemoglobin was 12.77±2.03 gm per cent and among female the mean was 10.38±1.17 gm per cent. Prevalence of anemia was found to be 63.12 per cent. Mild anemia (Hemoglobin = 10-11.9 gm per cent) was found to be 53.28 per cent. A total of 17.21 per cent had moderate degree anemia (Hb 7-9.9 gm per cent). Prevalence of severe anemia (<7 gm per cent) was found to be 3.28 per cent and polycythemia (HB > 15 mg per cent) was found 1.64 per cent and also the distribution of different anemic conditions like Microcytic Blood Picture, Microcytic Anemia, Iron Deficiency Anemia, and Polycythemia were assessed *in three different tribal communities. Prevalence of anemia was substantially high among the tribal populations.*

***Key words:** Anemia, Haemoglobin, tribal women, Microcytic Blood Picture, Microcytic Anemia, Iron Deficiency Anemia, Polycythemia.*

Introduction

Anemia is a major public health problem among the Indian Population especially in the rural and tribal belt. The present study have shown that the tribal people have varying degrees of anemia due to illiteracy, insufficient food intake, lack of access to health services, unhygienic personal habits, adverse cultural practices and frequent infections. The people of these three tribes are predominantly depending on the agricultural product. Around 55 per cent of adult populations are agricultural labour. They lack proper educational and health facilities. Prevalence of certain belief systems, in respect of health and nutrition, aggravate the situation. Anemia is a major cause of high incidence of premature births, low birth weight, perinatal mortality, and maternal mortality. Hence the primary objectives of this study is to determine the distribution of Haemoglobin concentration among three tribes in different age group and to study the determinants of anemia among the three tribal population.

Subjects and Methods

An Anthropological fieldwork was conducted in Mayurbhanj district, which has been inhabited by multiple tribal communities in Odisha. Total population of the district is 2223456 and Scheduled Tribe (ST) is 57.67 per cent as per Census 2001. Literacy rate was 52.43 per cent, whereas male literacy was 66.38 per cent and female literacy was 38.28 per cent. Since the main objective of this study was to understand the anthropological dynamics with special reference to haematological parameters among multi-ethnic/ tribal communities, it was pertinent to select a location which is inhabited by multiple tribal communities. The place of our study is Shamakhunta Block of Mayurbhanj District of Odisha. Three tribes namely, Bhumija, Bathudi, and Sabara were selected for Hematological study. As Balidiha Panchayat in Mayurbhanj District is inhabited by several tribal communities such as Bathudi, Bhumija, Sabara, Santal, Kolha, Bhuinya, etc. it was decided to select the area keeping in view the practical aspects of conducting a field work involving diverse research interests of participation and operational aspects. After the completion of household census, the data was analysed to identify age, sex and marital status-wise distribution of subjects. Subjects with special demographic characteristics like those below five years of age, old, seriously ill, pregnant women, mentally disorder were excluded from the sampling. It was decided randomly to choose subjects those who consented to participate voluntarily in the sample collection. 2-5 ml of blood were collected from each individual. The total number of 122 samples were collected from three tribal communities. Tribe and sex-wise distribution of sample is presented in Table 8.1.

Table 8.1. Tribe and sex-wise distribution of sample

Tribe	*Male*	*Female*	*Total*
Bhumija	19	34	53
Bathudi	18	15	33
Sabara	17	19	36
Total	**54**	**68**	**122**

Subject participation and consent procedure - The villagers were explained about the objectives, methods and expected outcome before the collection of blood sample. To avoid the risk of procedural difficulties a trained para-medical staff having professional expertise was requested to be engaged to assist in the collection of blood sample. In the morning of the day of data collection a makeshift camp was established in the village where all the subject participants were asked to give their blood samples with proper consent. Standard ethical protocols in accordance to the Indian Council of Medical Research (ICMR) were followed in the collection of blood sample. Maximum care was taken in informing the villagers about the purpose of the study before the sample collection. Individual subject participants were further explained about the study and the possible utilities of their sample and then written consent was taken from them. In case of illiterate participants, their consent was taken after making them understand the whole dynamics of the study.

Transportation and Analysis - In order to avoid the damage of the blood cells kept in the EDTA vile, arrangements were made to transport the samples from the field site to the Biological Laboratory at the Department of Anthropology, Utkal University, Bhubaneswar, which is 280 Kms from the field work site and takes a seven hours long journey. The analysis was done with the help of Sysmax KX 21 analyser. After reaching the laboratory the samples were again codified with new identifying marks to maintain highest standard of confidentiality and to protect the identity of subject. The haematological parameters were analysed by the Sysmax KX 21 analyser in the presence of the expert research scholars and supervisors. Primary outcome variables in the analysis were created from hemoglobin measurements. We classified women as mildly, moderately or severely anemic based upon their hemoglobin status and World Health Organisation (WHO) classification. A hemoglobin concentration of less than 7 g/dl was used to define severe anemia, 7-9.9 g/dl for moderate anemia, 10-12 g/dl for mild anemia and more than 16 g/dl for polycythemia. Analysing the MCV (Mean Corpuscular Volume), MCH (Mean Corpuscular Haemoglobin), MCHC (Mean Corpuscular Haemoglobin Concentration) and HGB (Haemoglobin), the classification of Anemia is drawn like as MBP = Microcytic Blood Picture (MCV Less), MA = Microcytic Anemia (Less HGB, MCV), IDA = Iron Deficiency Anemia (Less MCV, MCH, MCHC, HGB), Anemic (Less HGB), and Polycythemia (More HGB) on the basis of normal cut off range of HGB, MCV, MCH, and MCHC (by WHO). The data were analysed using Statistical Package for the Social Sciences (SPSS) version 20 software.

Sharing of result with the subject participants–A copy of the report of the blood sample analysis was compiled and then handed over to the concerned subject by the researcher after physically visiting his/her household.

Decomposing the Bio-Degradable Waste Materials–The entire disposable syringe, EDTA vile and needles are properly collected after the inspection and proper procedure maintained as per ICMR guidelines.

Results

The level of heamoglobin among individuals of different age groups of three tribes of Odisha is evident from Table 8.2. The Bhumija of all ages have heamoglobin 10.88 g/dl and for male it was 11.74 g/dl whereas for female it was found 10.40 g/dl. Among Bathudi tribes of all ages, it was 10.89 g/dl; which varies from 11.64 g/dl for males to 9.98 g/dl for females. Mean of heamoglobin level among Sabara of all ages was found 11.51 g/dl and for male it is 12.77 whereas for female it is 10.38. The F value does not show a significant statistical difference in terms of mean Haemoglobin (HGB g/dl) values among three tribes.

Table 8.2: Age and sex wise mean Haemoglobin (HGB g/dl) among three tribes of Odisha

AGE GROUP	*SEX*	*Bhumija*			*Bathudi*			*Sabara*		
		N	*Mean*	*SD*	*N*	*Mean*	*SD*	*N*	*Mean*	*SD*
< 10	Male	00	-	-	00	-	-	00	-	-
	Female	01	10.7	.	00	-	-	00	-	-
11-20	Male	01	11.9	.	02	11.95	0.78	02	14.75	2.62
	Female	10	10.89	0.68	04	11.15	1.53	01	11.2	.
21-30	Male	05	11.82	1.79	03	13.57	0.84	03	14.77	1.7
	Female	09	10.8	0.67	03	9.93	1.31	01	13.2	.
31-40	Male	03	11.4	0.46	02	13.05	2.47	03	12.9	1.22
	Female	03	8.33	4.01	03	10.8	0.66	05	10.3	0.74
41-50	Male	02	12.3	0.85	05	11.54	0.86	05	11.36	1.94
	Female	05	10.56	0.83	02	8.35	3.32	04	10	1.61
51-60	Male	06	11.73	0.91	02	11.25	0.64	01	13.4	.
	Female	04	10.48	1.48	01	8.8	.	03	10.53	1.01
60 +	Male	02	11.45	1.63	04	9.68	2.46	03	11.47	0.35
	Female	02	8.6	2.12	02	8.7	0.28	05	9.96	0.78
Total	Male	19	11.74	1.1	18	11.64	1.89	17	12.77	2.03
	Female	34	10.4	1.53	15	9.98	1.69	19	10.38	1.17
	Total	**53**	**10.88**	**1.53**	**33**	**10.89**	**1.96**	**36**	**11.51**	**2.01**

($F= 1.533$; $df =2$, $p = 0.220$)

Table 8.3 shows the distribution of tribes as per level of aneamia. Mild anemia (Hemoglobin = 10-11.9 gm per cent) was found to be 53.28 per cent. A total of 17.21 per cent had moderate degree anemia (Hb 7-9.9 gm per cent). Prevalence of severe anemia (<7 gm per cent) was found to be 3.28 per cent and polycythemia (HB > 15 mg per cent) was found 1.64 per cent.

Table 8.3: Distribution of studied tribe as per level of anemia

Anemia Condition / *Different studies on Tribes*		*Normal (Hb=12-14.9 gm per cent) N (per cent)*	*Mild anemia (Hb= 10-11.9 gm per cent) N (per cent)*	*Moderate degree anemia (Hb = 7- 9.9 gm per cent) N (per cent)*	*Prevalence of severe anemia (HB <7 gm per cent) N (per cent)*	*Polycythemia (Hb > 15 gm per cent) N (per cent)*
Results in present study	Bhumija (N=53)	11 (20.75)	34 (64.15)	07 (13.21)	01 (1.89)	00 (00)
	Bathudi (N=33)	09 (27.27)	15 (45.45)	07 (21.21)	02 (06.06)	00 (00)
	Sabara (N=36)	10 (27.78)	16 (44.44)	08 (22.22)	00 (00)	02 (5.56)
	Total (N= 122)	**30 (24.59)**	**65 (53.28)**	**22 (18.03)**	**O3 (2.46)**	**02 (1.64)**
Kharia Tribes of Sundargarh District (N=638) R. S. Balgir, et.al., 2003		20 (10.1)	107 (54.0)	68 (34.3)	3 (1.5)	NA
Bhuyan Tribes of Sundargarh District (N=198) R. S. Balgir, et.al., 2003		167 (26.2)	346 (54.2)	121 (19.0)	4 (0.6)	NA
Tribals in Wayanad district of Kerala N=347 Shrinivasa BM, et.al. 2014		12 (3.5)	106 (30.5)	194 (55.9)	35 (10.1)	NA

Anemia Condition	*Results in present study N= 122*	*Kharia Tribes of Sundargarh District (N=638) R. S. Balgir, et.al., 2003*	*Bhuyan Tribes of Sundargarh District (N=198) R. S. Balgir, et.al., 2003*	*tribals in Wayanad district of Kerala N=347 Shrinivasa BM, et.al. 2014N*
	N (per cent)	*N (per cent)*	*N (per cent)*	*(per cent)*
Normal (hemoglobin=12-14.9 gm per cent)	30 (24.59)	20 (10.1)	167 (26.2)	12 (3.5)
Mild anemia (Hemoglobin = Hb 10-11.9 gm per cent)	65 (53.28)	107 (54.0)	346 (54.2)	106 (30.5)
Moderate degree anemia (Hb 7-9.9 gm per cent)	21 (17.21)	68 (34.3)	121 (19.0)	194 (55.9)
Prevalence of severe anemia (HB <7 gm per cent)	O4 (3.28)	3 (1.5)	4 (0.6)	35 (10.1)
Polycythemia (hemoglobin > 15gm per cent)	02 (1.64)	NA	NA	NA

Table 8.4: Distribution of studied tribes as per different anemic condition

Anemic Condition	*Tribe*						*Total*	
	Bhumija		*Bathudi*		*Sabara*			
	N	*per cent*	*N*	*per cent*	*N*	*per cent*	*N*	*per cent*
Normal	10	18.87	07	21.21	13	36.11	30	24.59
Microcytic Blood Picture	26	49.06	13	39.39	02	5.56	41	33.61
Microcytic Anemia	00	0.00	00	0.00	03	8.33	03	2.46
Iron Deficiency Anemia	12	22.64	05	15.15	09	25.00	26	21.31
Anemic	00	0.00	04	12.12	01	2.78	05	4.10
Polycythemia	00	0.00	00	0.00	02	5.56	02	1.64
Others	05	9.43	04	12.12	06	16.67	15	12.30
Total	**53**	**100**	**33**	**100**	**36**	**100**	**122**	**100**

Distribution of studied tribes as per different anemic condition is presented in Table 8.4. It is apparent that among Bhumija tribe most of the people (49.06 per cent) were having Microcytic Blood Picture remaining 18.87 per cent were in normal condition, 22.64 had Iron deficiency Anemia and 9.43 were having other diseases like Eosinophilia, platelet deficiency etc. Similarly, among Bathudi tribe most of the people were having Microcytic Blood Picture (39.39 per cent), remaining 21.21 per cent were in normal condition, 15.15 per cent were suffering from Iron deficiency, 12.12 per cent were Anemic and 12.12 per cent were having other diseases like Eosinophilia, platelet deficiency, Neutrophilia, Leukocytosis etc. Among Sabara Tribe, 36.11 per cent Samples were found as normal, 25 per cent sample are having Iron Deficiency, 5.56 per cent were Microcytic Blood Picture, 8.33 per cent were microcytic Anemia, 2.72 per cent are anemic, 5.56 were having Polycythemia and remaining 16.67 per cent sample were suffering from other disorders like Eosinophilia, platelet deficiency, Neutrophilia, Leukocytosis etc.

It is apparent from the Table 8.5 that among Bhumija 71.7 per cent, among Bathudi 66.66 per cent and among Sabara 47.23 per cent of population were anemic. Further, as per pooled data, a total of 63.12 per cent tribes were anemic.

Table 8.5: Prevalence of anemia among the studied tribes

Anemic Condition	*Tribe* *Bhumija*		*Bathudi*		*Sabara*		*Total*	
	N	*per cent*	*N*	*per cent*	*N*	*per cent*	*N*	*per cent*
Anemic	38	71.7	22	66.66	17	47.23	77	63.12
Normal	15	28.3	11	33.33	19	52.78	45	36.89
Total	**53**	**100**	**33**	**100**	**36**	**100**	**122**	**100**

Table 8.6. Distribution of studied tribes as per different type of anemia

Type of anemia/Disease	*Tribe* *Bhumija*		*Bathudi*		*Sabara*		*Total*	
	N	*per cent*	*N*	*per cent*	*N*	*per cent*	*N*	*per cent*
Normal	10	18.87	06	18.18	13	36.11	29	23.77
Microcytic Blood Picture	14	26.42	06	18.18	02	5.56	22	18.03
Microcytic Blood Picture and Eosinophilia	11	20.75	06	18.18	00	0.00	17	13.93
Microcytic Blood Picture and Neutrophilia	01	1.89	01	3.03	00	0.00	02	1.64
Microcytic Anemia	00	0.00	00	0.00	03	8.33	03	2.46
Iron Deficiency Anemia	10	18.87	03	9.09	08	22.22	21	17.21
Iron Deficiency Anemia and Eosinophilia	02	3.77	02	6.06	01	2.78	05	4.10
Anemic	00	0.00	02	6.06	01	2.78	03	2.46
Anemic and Eosinophilia	00	0.00	01	3.03	00	0.00	01	0.82
Anemic, Esoinophilia and Neutrophilia	00	0.00	01	3.03	00	0.00	01	0.82
Polycythemia	00	0.00	00	0.00	02	5.56	02	1.64
Eosinophilia	04	7.55	03	9.09	06	16.67	13	10.66
Leukocytosis	00	0.00	01	3.03	00	0.00	01	0.82
Platelet deficiency	01	1.89	01	3.03	00	0.00	02	1.64
Total	**53**	**100**	**33**	**100**	**36**	**100**	**122**	**100**

Distribution of studied tribes as per different type of anemia is presented in Table 8.6. It is apparent that all the three tribes i.e. Bhumija, Bhatudi and Sabara having Microcytic Blood Picture, Microcytic Blood Picture with Eosinophilia, anemia, Iron deficiency Anemia, Iron deficiency Anemia with Eosinophilia, Anemic, Eosinophilia and Neutrophilia, Eosinophilia, platelet deficiency etc.

Conclusion

The low socio-economic status (SES) of tribal people force them to work in low-income jobs, resulting in poor purchasing power, hence low quality diet, Undernutrition and anemia. The present study revealed that most of the people are working as daily wage labourers who do not avail good health facilities, live in unhygienic atmosphere and non-sanitary facilities. It also highlights that all these factors aggregate to bring abnormal haematological profile among the studied communities which is a major health concern. It also shows that there are differences that vary among the three tribes with respect to haematological parameters. It could be concluded that the Sabara people were more anemic than Bhumija and Bathudi.

References

Balgir, R. S. , R. K. Mishra and B.Murmu, 2003. "Clinical and Hematological Profile of Hemoglobinopathies inTwo Tribal Communities of Sundargarh District in Odisha, India." *Kamla-Raj* 3(4): 209-216.

Shrinivasa BM, Rekha Rachel Philip, Vijaya Kumar Krishnapali, Asha Suraj, PR Sreelakshmi, AprJun 2014. "Prevalence of anemia among tribal women of reproductive agegroup in Wayanad district of Kerala", *International Journal of Health & Allied Sciences*, Vol. 3, Issue 2.

Indian Council of Medical Research 1989. "Growth and physical development of Indian infants and children." New Delhi.*Tech. Report Series No.18. ICMR.*

Rao, S. S., Goyal, J. P., Raghunath, S. V., & Shah, V. B. 2012. "Hematological profile of sickle cell disease from South Gujarat, India." *Hematology reports*,4(2).

Wojtyla C., Biliñski1, P., Paprzycki P., Warzocha1 K., 2011 "Haematological parameters in postpartum women and their babies in Poland –comparison of urban and rural areas," *Annals of Agricultural and Environmental Medicine*,Vol 18, No 2, 380385.

Mohapatra M. 2013. "A Comparative Study on the Haematological Status of Bhumija and Sabara Communities of Balidiha, Mayurbhanj, Odisha," unpublished MSc Dissertation Submitted to P.G. Department of Anthropology, Utkal University.

Theme – 3
Adult Nutrition, BMI and Puberty

9

Age at Puberty and Nutritional Status: A Comparative Study among Adult Females of two Different Ecological Zones

—Rachna Thakur and Rajesh K. Gautam

Abstract

Age at menarche is a sign of puberty among women. It is detemined by bio-social characteristics, viz. socio-economic condition, culture, ecology etc. whereas body mass index (BMI) is a scale to assess the level of nutrition. In the present study women of different ecological setup as well as culture were studied for their onset of age at menarche and level of nutrition. Cross-sectional data was gathered on a total of 446 women aged 18-60 years from six different endogamous groups of two ecological zones. The mean age at menarche varies from 13.66 years among Brahmins of Madhya Pradesh to 16.23 years among the Rajput of Uttarakhand; Similarly, variation in mean BMI ranging between 18.56 Kg/m² and 20.71 Kg/m² among these women. It was found that altitude is one of the determinants of the age at menarche. Similarly, socio-economic statuses also have impact on puberty of children. On the basis of present study it can be postulated that the nutritional status of Indian women is no doubt improved during last two decades. But still large numbers of women are malnourished especially in deprived section of the societies and require special attention.

Introduction

Women are better half of any society and country, they constitute half of the Indian population their health has major concern, because it has a direct concern to the rest of the population. In the present study, an attempt was made to find out biosocial correlates of age at menarche and nutritional status among adult females of rural India.

Menarche is an important event in the life of a girl child. Numerous factors including genetic influences, socioeconomic condition, culture ecology, general health and wellbeing nutritional status and some type of exercise influence the age at menarche. A correct appraisal of the age of menarche of our population is a pressing need, without which no social law, especially those related to marriage can be made. A shift in the age at menarche signifies a concurrent change in the factors associated with it. Menarche is celebrated in many cultures around the world as rites of passage, a time to recognize that a girl is moving into or has attained womanhood. There are many scholars paid attention on age at onset of menarche, some of them are Adak *et al.*, (2006a) and (2006b), Gautam *et al.*, (2008), Kshatriya *et al.*, (2009) and Kumar and Gautam (2014).

Anthropometry, despite its inherent limitations, still remains the most practical tool for assessing the nutritional status in the community (Ghosh *et al.*, 2001). Body mass index (BMI) expressed as a ratio of weight to height square can be a good parameter to assess the nutritional level and chronic energy deficiency (CED) in adults (Naidu, Neela & Rao, 1991). There are many studies based on this aspect (Ferro-Luzzi *et al.*, 1992; Naidu & Rao, 1994; Khongsdier, 2001, Gautam *et al.*, 2006 and Gautam 2009). But, the correlation of BMI and age at menarche is least explored, hence there is an attempt to study the level of nutrition and age at menarche among the women of two different ecology.

Methodology

The study sample was drawn from two geo-ecological regions i.e. Plains of Madhya Pradesh and Himalayan Hills of Uttrakhand state of India. A multistage stratified sampling technique was adopted for selection of sample. After selection of states, one district was selected from each of the states. In this way district Sagar was selected from

Madhya Pradesh and district Nainital from Uttarakhand; again five and eight villages were selected respectively from both the districts. Six endogamous groups viz. 1. Brahmin, 2.Thakur, 3.Chamar (of Madhya Pradesh); and 4. Brahmin, 5. Rajput and 6. Arya (of Uttarakhand) were selected for the present study. The criteria for inclusion/exclusion of sample were as follows:

1. Women between the ages of 18 to 60 years.
2. Not currently pregnant.
3. Had not given birth during the last six months.
4. Had not experienced fetal loss viz. abortion or still birth during the last three months.
5. Was not severely ill during the last three months.
6. Not menstruating at time of study.
7. Not with any congenital malformation of any degree.

Finally the women were selected randomly. Total 446 women were selected for the study. Efforts were also made to exclude closely-related individuals like sisters, mothers and daughters and those with any kind of physical enormities. Therefore, the sample was free of selection bias.

In this study, body weight (W) and height (H) was measured and BMI (W/H2) was calculated for each individual followed by calculation of central tendency viz. mean and deviation of each of the measurements among each population.

Area and People

Among the world's ancient and the greatest civilisation, Indian culture and civilisation is unique. It has spread over 3287263 square km. India is the world seventh largest country. In term of population it ranked second. It is rich in biodiversity and diverse in natural resources.

Madhya Pradesh: Sagar

Madhya Pradesh is the second largest state in the country. Geographically, it has important place in our country. It is characterised by low hills, extensive plateaus, and river valleys. The district of Sagar lies in the north-central region of Madhya Pradesh. The total area of the district is 10,252 sq. kms. It is the 16th largest district in the state. It is inhabited by people of diverse ethnic, religious and linguistic backgrounds. Hindi is the main language spoken in the district (Bhatt, 1998). The sample for the present study was taken from the following three endogamous caste groups of five different villages of the district.

1. Brahmin: The Brahmins are priestly caste of India and are at the top of the caste hierarchy of Hinduism. They are also posses' large agricultural land who do not practice agriculture but instead hire labour for agricultural tasks. Some of them in service sector and business also. A total of 50 women of Brahmin caste were included in the present investigation. The women do not enjoy equal privilege like the males and widespread sex discrimination can be observed. The sample was drawn from rural families who are considered more dogmatic.

2. Thakur: They consider themselves as the warrior class. The warriors occupy the second position in the Hindu caste hierarchy and are known as the *Kshatriya.* They subsist on agriculture and animal domestication. Very few of them are in other sectors. For the present investigation, a total of 100 Thakur women were chosen.

3. Chamar or Ahirwar: In the vertical caste hierarchy of Hinduism, the Chamar are at the bottom. They are considered 'untouchables' especially in rural India. There is strong prohibition against the Chamar in some jobs. If they are employed, their identity is concealed. In order to provide constitutional protection, Chamar is included in the list of scheduled castes who suffer from similar discrimination. A total of 34 women of this caste were included in the present investigation.

Uttarakhand: Nainital

Uttarakhand is described in ancient Hindu religious text as Kedarkhand, Manaskhand and Himbandh. Due to its holy and pilgrimage places, it often called- Devbhumi. It is a hilly state; spread on the Siwalik range of the Himalayan foothills. Geographically Nainital district is divided into two zones viz. Hilly and Bhabar. The samples for present study were drawn from eight different villages located in the Hilly region of the district which is above 1500 meters mean sea level (Bhatt, 1998). Brahmin, Rajput and Arya were the three endogamous caste groups selected for inclusion in the present study. Brief details of each of these castes are given below:

1. *Brahmin:* The Brahmin of Uttarakhand is slightly different from the Brahmin of Madhya Pradesh. They do not practice priesthood. Most of them are non-vegetarians. They practice agriculture and animal domestication. In social ranking they are at the top, like other Brahmins in India. Although they enjoy higher social status, they have to work in tough ecological conditions where their terraces like agricultural fields are located. A total of 60 Brahmin women were selected for the study.
2. *Rajput:* Uttarakhand is dominated by the Rajputs. They practice agriculture and horticulture; simultaneously they domesticate the cow and buffalo. Beside traditional occupations, they are also employed in the Indian army in large numbers and hold most of the government jobs and political positions. A total of 150 Rajput women of Uttarakhand living in five villages were included in the present investigation.
3. *Arya:* Arya is a group of artisans that includes blacksmiths, coppersmiths, carpenters and weavers. These artisans are considered as 'untouchables' and fall into the scheduled castes; they are provided constitutional safeguards. Still, they are treated as 'unclean' and are prohibited in many day-to-day activities. A total of 52 Arya women were included in the present study.

Results

The descriptive statistics of age at menarche among the studied population are presented in Table 9.1. It is evident from the Table that the mean age at menarche is higher among the women of Uttarakhand as compared to Madhya Pradesh, further it is lowest (13.6 years) for Brahmin women of Madhya Pradesh and highest (16.1) among the Rajput of Uttarakhand. Among the present studied women, they at menarche varies from 11 to 22 years. For further elucidation of facts box plot diagram constructed which clearly indicates the mean and extent of age at menarche in different population of two ecological zones (Figure 9.1).

Table 9.1: Central tendencies, standard deviation and standard error of age at menarche among the women of Madhya Pradesh and Uttarakhand

Age at Menarche	*Rajput (UK)*	*Brahmin (UK)*	*Arya (UK)*	*Thakur (MP)*	*Ahirwar (MP)*	*Brahmin (MP)*
Sample size	149	59	52	99	34	50
Mean	16.13	15.61	15.29	14.95	14.44	13.66
S.E.	.133	.189	.204	.137	0.18	0.13
Median	16.00	16.00	15.00	15.00	14.0	14.0
Mode	16	16	15	14	14	14
S.D.	1.620	1.451	1.473	1.366	1.05	0.93
Minimum	12	13	11	12	13	12
Maximum	22	20	18	20	18	15

Caste-wise descriptive statistics of body mass index (BMI) is presented in Table 9.2. It was found that mean BMI was 19.5 kg/m^2 for pooled data. Table reveals that the mean BMI of Rajput (UK) women of Uttarakhand was 19.4 kg/m^2; whereas it was found 18.56 kg/m^2 for Brahmin (UK) and 19.33 kg/m^2 among Arya (UK). At the same time the women of Madhya Pradesh namely, Thakur (MP) have mean BMI 19.94 kg/m^2; Ahirwar (MP) have 19.22 kg/m^2; whereas the Brahmin (MP) women of the state have higher mean BMI (20.70 kg/m^2) among

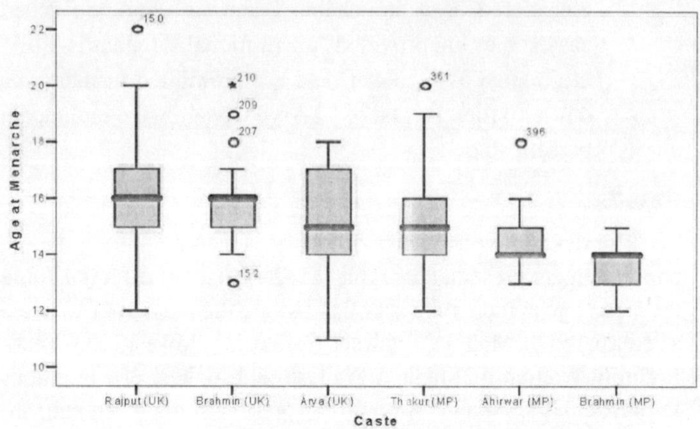

Figure 9.1: Box plot diagram showing Central Tendency and dispersion of age at menarche among different population of two ecological zones.

all the studied caste groups. For further elucidation bar diagram constructed (Figure 9.2), which clearly shows that Brahmin women Uttarakhand have lowest mean BMI and the Brahmin women of Madhya Pradesh have highest mean BMI among the six caste group studied.

Table 9.2: Caste wise Central Tendency and dispersion of Body Mass Index among woman of two ecological zones

BMI	*Rajput (UK)*	*Brahmin (UK)*	*Arya (UK)*	*Thakur (M.P.)*	*Ahirwar (MP)*	*Brahmin (MP)*
N	150	60	52	100	34	50
Mean	19.43	18.57	19.83	19.95	19.20	20.71
S.E.	0.26	0.33	0.53	0.36	0.44	0.39
Median	19.28	17.92	18.90	19.18	19.05	20.53
Mode	18.43	16.66	16.23	16.44	17.48	18.18
S.D.	3.22	2.53	3.81	3.56	2.60	2.82
Minimum	12.31	14.38	12.80	13.87	15.38	15.22
Maximum	33.32	26.38	31.47	32.59	27.77	26.67

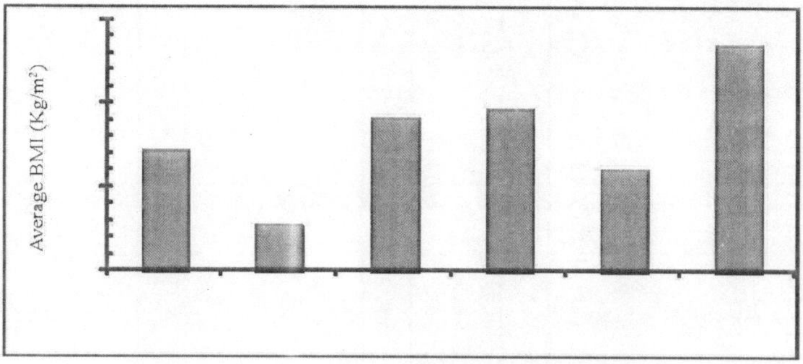

Figure 9.2: Bar diagram showing Caste-wise mean Body Mass Index.

Caste-wise nutritional statuses of studied women were presented in Table 9.3. It is evident from the Table that 42.1 per cent women were lean and thin (fell in CED grade), 50.7 per cent have normal nutritional status and 7.2 per cent women were obese. Further Brahmin women of Uttarakhand have poor nutritional status as 58.3 per cent of them were found chronic energy deficient followed by Arya (44.2 per cent) and Rajput (41.3 per cent). On the other hand among the Brahmin women of Madhya Pradesh only 14 per cent were found chronic energy

Table 9.3. Distribution of population as per level of Nutrition (CED Grade)

Population	*Level of Nutrition*														*Total*	
	CED Grade III (Severe) <16.0		*CED Grade II (Moderate) 16.0-16.99*		*CED Grade I (Mild) 17.0-18.49*		*Low Weight Normal (18.5-19.99)*		*Normal (20.0-24.9)*		*Obese Grade I (25.0-29.99)*		*Obese Grade II (>30)*			
	N	*per cent*	*N*	*per cent*	*N*	*per cent*	*N*	*per cent*	*N*	*per cent*	*N*	*per cent*	*N*	*per cent*	*N*	*per cent*
Rajput (UK)	20	13.3	15	10.0	27	18.0	36	24.0	44	29.3	7	4.7	1	0.7	150	100
Brahmin (UK)	8	13.3	8	13.3	19	31.7	10	16.7	14	23.3	1	1.7	0	-	60	100
Arya (UK)	2	3.8	11	21.2	10	19.2	8	15.4	16	30.8	3	5.8	2	3.8	52	100
Thakur (MP)	5	5.0	14	14.0	21	21.0	24	24.0	24	24.0	10	10.0	2	2.0	100	100
Ahirwar (MP)	2	5.9	3	8.8	11	32.4	9	26.5	7	20.6	2	5.9	0	-	34	100
Brahmin (MP)	1	2.0	3	6.0	8	16.0	11	22.0	23	46.0	4	8.0	0	-	50	100
Total	**38**	**8.5**	**54**	**12.1**	**96**	**21.5**	**98**	**22.0**	**128**	**28.7**	**27**	**6.1**	**5**	**1.1**	**446**	**100**

deficient whereas a total of 48 per cent of Thakur (MP), 47.1 per cent of Ahirwar (MP) women were found chronic energy deficient. For further elucidation of facts the distribution of women of different caste group as per nutritional status are shown in a bar diagram (Figure 9.3).

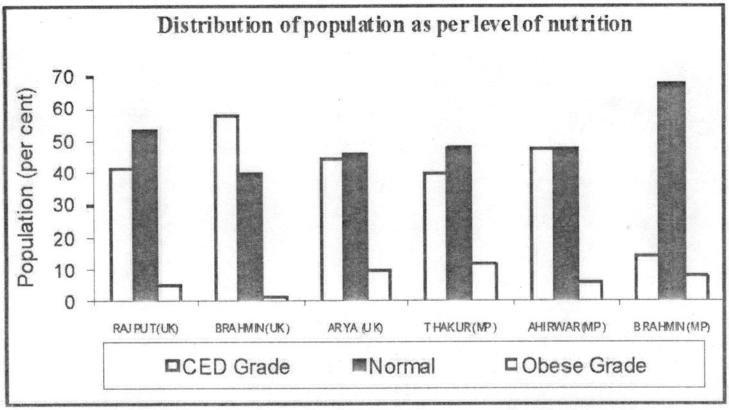

Figure 9.3: Bar diagram showing Caste-wise CED Grade.

Discussion

In the present study a total of 42 per cent women were found chronic energy deficient which higher as compared to the finding of National Family Health Survey (NFHS-3), according to which 36 per cent of women have a BMI below 18.5. Further, according to NFHS-3, women suffer from a dual burden of malnutrition with nearly half of them being either too thin or overweight; the per cent of ever-married women in the 15-49 years age–group who are overweight or obese increased from 11 per cent in NFHS-2 (National Family Health Survey) to 15 per cent in NFHS-3. In present study a total of 7.2 per cent women were found overweight.

To understand the correlation of age at menarche and BMI a scattered plot diagram is constructed for women aged 18-25 years of age, which is presented as Figure 9.4. It is apparent that there is almost no correlation in age at onset of menarche and BMI. The reason behind such correlation is current age of present sample. In this study, the sample women were from 18-60 years of age group, although the scattered plot diagram is constructed from the women aged 18-25 years of age group, still they are beyond the purview of age at occurence of puberty.

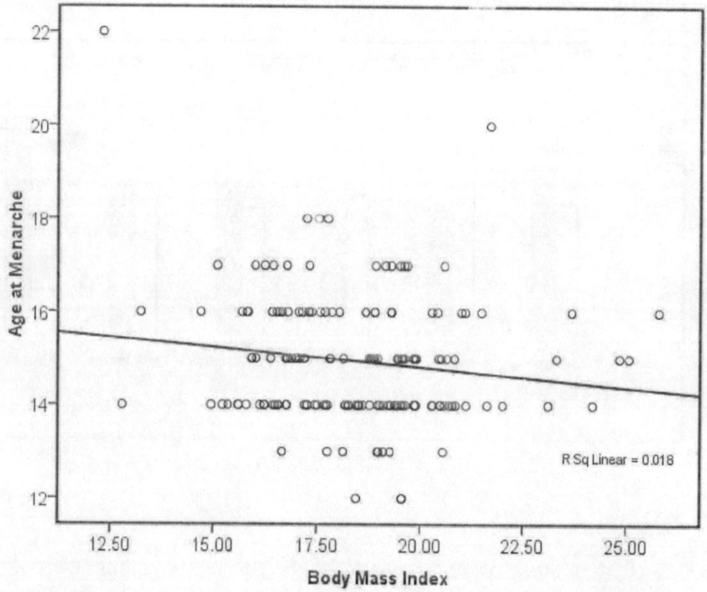

Figure 9.4: Scattered plot diagram showing correlation between BMI and age at menarche among women of 18 to 25 years of age.

A comparison of mean BMI of Indian women from previous studies (Ferro-Luzzi *et al.*, 1992; NFHS-III, 2007) has led to the conclusion that the nutritional status of Indian women had improved over the last two decades. This has taken place because of the overall growth of socio-economic condition of the country, women empowerment, rural employment and improved educational attainment of women. The outcomes of welfare schemes launched after independence are now perceptible. The improved nutritional status of the studied women can be seen as outcomes of family welfare, maternal and child health and reproductive and child health programmes (RCH) launched all over the country.

Conclusion

This study leads to conclude that the women of Uttarakhand have delayed menarche as compared to the women of Madhya Pradesh. This could be primarily because of difference in their ecology. The women of Uttarakhand are living in middle altitude 1500 meter from mean sea level, which is trans-Himalayan region; whereas the women of Madhya Pradesh are residing in the plains, below 1000 meter from mean sea level. Wast majority of women (42 per cent) were found lean and thin. They have BMI <18.5 kg/m^2. Overweight and obesity is also witnessed among these women, although in little proportion (seven per cent); but it is an indication towards nutritional transition. There is dual burden of the problem. There is need to proper implementation of supplementary nutritional programme being run for pregnant and nutritional programme. Beyond that great emphasis is needed to eradicate undernutrition from childhood as the chronic energy deficiency at adulthood is an outcome of stunting, wasting and undernutrition at childhood and high prevalence of childhood undernutrition is being reported from different corner of the country (Mondel and Sen, 2010a and 2010b, Thakur, R and Gautam, RK 2015a, 2015b; and 2015c; Ahirwar and Gautam 2015). Certainly, there would be correlation between BMI and age at onset of puberty, but this study has failed to establish it because of most of the sample were of >18 years of age.

References

Adak, DK, Gautam RK and Gharami AK (2006b). Assessment of Nutritional status through Body Mass Index among adult males of 7 tribal populations of Maharashtra, India. *Mal J Nutr* 12 (1): 23-31.2006.

Adak, DK, Gautam RK, Bharati S, Gharami AK, Pal M and Bharati P (2006a). Body Mass Index and chronic energy deficiency of adult males of Central Indian populations. *Human Biology* 78 (2): 201–218.

Ahirwar A and Gautam RK. 2015. Growth pattern among Bharia boys- A tribe of Patalkot, Chhindwara (MP), India. Human Biology Review, 4 (3), 221-235.

Bhatt SC (1998). The Encyclopaedic District Gazetteers of India, Gyan Publishing House, New Delhi. Vol. 6, 885-891.

Ferro-Luzzi A, Sette S, Franklin M and James WPT (1992). A simplified approach of assessing adult chronic energy deficiency. *Eur J Clin Nutr* 46:173-186.

Frankenfield, DC, Rowe WA, Cooney RN, Smith JS and Becker D (2001). Limits of body mass index to detect obesity and predict body composition. *Nutrition* 17: 26-30.

Gautam, RK, Thakur R. 2009. Biosocial correlates of Nutrition and chronic energy deficiency among adult females of two ecological zones in Madhya Pradesh and Uttarakhand, India. *Mal. J. Nutr* 15(2): 137-153.

Gautam, RK and Adak DK (2007). Population, health and nutrition in Central India: a situational analysis. *Mal J Nutr* 13 (1):55-70.

Gautam, RK (2007). Biosocial covariates of adult male body mass index in Central India. *J Biosoc Sci* 39: 875-893.

Gautam, RK (2008). Traditional occupation and nutritional adaptation among Central Indian caste populations. *J Biosoc Sci* 40: 697-723.

Gautam, RK, Adak DK, Gharami AK and Datta T (2006). Body Mass Index in Central India: inter-district variation. *Anthropologischer Anzeiger*: 64(4): 447-461.

Gordon, JE, Ascoli W, Mata LJ, Guzman MA and Scrimshaw NS (1968). Nutrition and infection field study in Guatemalan villages 1959-1964. Acute diarrheal disease and nutrition disorder in general disease incidence. *Archives of Environmental Health* 16: 424-437.

Gurrici, S, Hartriyanti Y, Hautvast JG and Deurenberg P (1998). Relationship between body fat and body mass index: Healthy percentage body fat ranges: an approach for developing guidelines based on body mass index. *Am J Clin Nutr* 72: 694 –701.

James, WPT, Ferro-Luzzi A and Waterlow JC (1988). Definition of chronic energy deficiency in adults. *Eur J Clin Nutr* 42: 969-981.

Khongsdier, R (2001). Body mass index of adult males in 12 populations of Northeast India. *Ann Hum Biol* 28: 374-383.

Kumar, A., Gautam, R. K. (2015). Obesity and chronic deficiency among adult bidi workers of Central India. *International Journal of Interdisciplinary and Multidisciplinary,* 2(5), 60–69.

Kyle, UG, Schutz Y, Dupertuis YM and Pichard C (2003). Body composition interpretation: contribution of the fat free mass index and the body fat mass index. *Nutrition* 19: 597-604.

Martin, R and Saller K (1956). *Lehrbuch der Anthropologie*, Vol. 3, Stuttgart, G. Fisher.

Miller, FJ, Billewicz WZ, Thomson AM (1972). Growth from birth to adult life of 442 Newcastle-upon-Tyne children. *Br J Prev Soc Med* 26: 224–230.

Mitra, A (1985). The nutrition situation in India. In: *Nutrition and Development*. M. Biswas and P. Andersen. (eds). Oxford, Oxford University Press, pp. 142-162.

Mondal, N, and Sen J. 2010a. Prevalence of undernutrition among children (5–12 years) belonging to three communities residing in a similar habitat in North Bengal, India Annals of Human Biology.

Mondal, N, and Sen J. 2010b. Prevalence of stunting and thinness among rural adolescents of Darjeeling district, West Bengal, India JPH - Year 8, Volume 7, Number 1, 2010.

Naidu, AN and Rao NP (1994). Body mass index: a measure of nutritional situation in Indian populations. *Eur J Clin Nutr* 48 (Suppl. 3) : 131-140.

Naidu, AN, Neela J and Rao NP (1991). Maternal body mass index and birth weight. *Nutrition News,* National Institute of Nutrition, Hyderabad 12, 2. *Biosocial Correlates of Nutrition and Chronic Energy Deficiency among Adult Females* 153.

NFHS-III (2007). National Family Health Survey (NFHS-3) 2005-06. International Institute of Population Sciences, Deonar, Mumbai.

Norgan, NG (1994). Relative sitting height and the interpretation of body mass index. *Ann Hum Biol* 21: 79-82.

Osmani, SR (eds.) (1992). Nutrition and Poverty. Clarendon Press, Oxford.

Thakur, R and Gautam RK 2015a. Prevalence of undernutrition among School going boys (5-18 years) of a Central Indian city (Sagar). *Human Biology Review*, 3 (4), 364-383.

Thakur, R and Gautam RK 2015b. Nutritional status among boys and girls of a central Indian Town (Sagar). Anthropological Review, Vol. 78 (2), 197–212.

Thakur, R and Gautam RK 2015c. Assessment of nutritional status among girls of 5-18 years of age of a Central Indian City (Sagar), *4*(4), 325–336.

10

Nutritional Status of Mawasis of Chhindwara District of Madhya Pradesh

—*Harshdeep Singh Dhanjal*

Abstract

Food is the basic need of all the organisms. Each and every organism requires food for sustenance. The foods we consume provide us energy which makes us capable to perform any work. Need of proper nutrition in life is necessary because it is the nutrition which keeps us healthy and strong and away from diseases. Food and nutrition are sometimes used synonymously but bear a lot of difference. Food is something edible which provide us nutrition whereas nutrition itself is the co-ordination of processes by which nutrients are consumed and utilised by the body. Hence, consumption of all types of food items, their digestion, absorption and assimilation in the body is referred as nutrition. In short, nutrition is the consumption of nutrients which are the chemical substances inside the food that nourish the body. All food items does not contain similar amount of nutrients but have nutrients of different types. Nutrition plays a vital role, as inadequate nutrition during childhood may lead to malnutrition, growth retardation, reduced work capacity and poor mental and social development .

Most of the people are aware of the food consumption and dietary practices but some are not. On an average, *two-third* of the world's population remains without food everyday. This leads to arise in some ill-health conditions in the body. When we talk about the consumption of food, we come over three terms, viz., Good nutrition, malnutrition and under nutrition. Under-nutrition continues to be a major problem especially in rural populations (Rao, 2001). But when malnutrition is concerned, it is still prevalent in developing countries. There are a lot of consequences of malnutrition among human populations living in different areas. There are some sort of dietary deficiencies in the food consumed by the people. These deficiencies negatively effect human health and economic progress of the country, leading to the important public health concerns and result in man-power and economic losses. Despite recent achievement in economic progress in India, the development has failed to secure a better nutritional status in children in the country. Thus, it is important for each and every individual to have better food and nutrition to live a comfortable and convenient life. There is dearth information on nutritional status of people particularly from tribal areas. Keeping this in view, the objective of present study was to assess the nutritional status of Mawasi tribal group residing in Kukarpani village in Junnardeo Development Block of Chhindwara district of Madhya Pradesh. An attempt is made to observe the nutritional status of the indigenous people. The key objective of this study is to calculate the BMI of the individuals and determine the status of their nutrition. Thus, this study is an attempt to throw light on the nutritional status of these tribals. These tribal groups do not possess any documentation regarding their nutrition. Thus, this study is an attempt to throw light on the nutritional status of these tribals.

Knowledge of nutritional status of community is necessary to have a comprehensive idea about its development process. Under-nutrition is one of the major health problems in developing countries. It is reported that the basic causes of undernutrition and infections in developing countries are poverty, poor hygienic conditions and little access to preventive healthcare (Mitra, 1985; WHO, 1990). Literature on BMI of adult Indians is limited to certain geographical areas or populations. Among them, are the study of BMI among the North-East Indians (Khongsdier, 2001), Central Indian (Gautam *et al.*, 2006, Adak *et al.*, 2006 a; b; Gautam and Adak, 2007, Gautam, 2007a; 2007b). Work on socio-economic and socio-cultural influences on women's nutritional status in India has been done by Chatterjee (1989). Studies on the tribal groups have also been done by various anthropologists. This includes work of Biswas and Kapoor (2003) on Mortality Among

Saharia – A Primitive Tribe of Madhya Pradesh. Similarly, a study on the nutritional status of 222 Khasi girls in Meghalaya was performed by Agrahar-Murugkar (2005). Joe, Mishra and Navaneetham (2009) worked on Childhood Malnutrition and published a paper named as 'Inequalities in Childhood Malnutrition in India: Some Evidence on Group Disparities.' Apart from this, some recent studies includes the work of Mallikharjuna Rao *et al.*, (2010) who worked on Diet and Nutritional status of Women in India. Inadequacies in nutritional intake or under-nutrition can be considered as a major source of many adverse effects on health of individuals (Gordon *et al.*, 1968). The critical review of the previous studies shows that there is no information about the Mawasi tribe. Hence there is an attempt to highlight their nutritional status.

Materials and Methods

For the present study data was collected from 104 individuals consisting of 53 males and 51 females. A total of 80 households were covered from the village Kukarpani. Anthropometric measurements such as height vertex and weight were taken for the calculation of Body Mass Index (BMI) of these individuals. The data was collected in a semi-structured schedule.

Anthropometry is considered to be an important tool for assessing nutritional status of individuals or of the community. Hence, measurements namely height and weight and calculation of BMI were done to assess the extent of malnutrition. Body mass index (BMI) expressed as a ratio of weight to height square can be a good parameter to grade chronic energy deficiency (CED) in adults (Naidu, Neela and Rao, 1991).

Small sample size is a limitation of present study and the sample were collected from a village only still it is first attempt to study the Mawasi tribe.

Results and Discussion

The description of mean height, weight and mean BMI of male and female population is given in Table 10.1. It is apparent from Table that the mean of height, weight and BMI among males was 165.58 cms, 54.28 kg and 19.79 Kg m^{-2} respectively; whereas females are shorter and lighter as compared to males. The mean height of females was 157.33 cms, mean weight was 46.41 kg and mean BMI was 18.78 Kg m^{-2}. Overall population depicts that the mean of height, weight and BMI among them is 161.54 cms, 50.42 kg and 19.29 Kg m^{-2} respectively.

Table 10.1: Mean and SD of Height, weight and BMI among Mawasi Tribe

S.No.	*Measurements*	*Male*	*Female*	*Total*
01.	Height (in cms.)	165.58(±18.38)	157.33(±20.78)	161.54(±18.38)
02.	Weight (in kg)	54.28(±13.24)	46.41(±7.14)	50.42(±13.24)
03.	Body Mass Index (BMI)	19.79(±10.87)	18.78(±1.83)	19.29(±10.87)

The distribution of BMI is presented in Table 10.2. Considering 18.5 as cut off point for screening individuals into normal and chronic energy deficient group, it was found that the prevalence of CED among males is 11.32 per cent. 43.40 per cent are low weight normal whereas only 45.28 per cent of the totals are normal. None of the person was found obese among the Mawasi.

Table 10.2: Percentage distribution of population as per Chronic Energy Deficiency level (male and female population)

Level of CED	*Male*		*Female*		*Total*	
	No.	*PC per cent*	*No.*	*PC per cent*	*No.*	*PC per cent*
CED Grade III Severe (< 16.0)	0	0	1	1.96	1	0.96
CED Grade II Moderate (16.0-16.99)	2	3.77	4	7.84	6	5.77
CED Grade I Mild (17.0-18.49)	4	7.55	16	31.37	20	19.23
Low Weight Normal (18.5-19.99)	23	43.40	21	41.18	44	42.30
Normal (20.0-24.9)	24	45.28	9	17.65	33	31.73
Obese Grade I (> 25.0)	0	0	0	0	0	0
Total	**53**	**100**	**51**	**100**	**104**	**100**

Female population exhibits a distinction from that of male population on the basis of CED level. 41.17 per cent of females fall under the category of CED which is 29.85 per cent more than males. 41.18 per cent of the females are low weight normal whereas only 17.65 per cent of them are normal. Like males, none of the females were found obese.

Total population shows that 25.96 per cent of the overall population (including both males and females) fall under CED level. Low weight individuals are 42.31 per cent in the population whereas 31.73 per cent of them are normal. None of the individuals are found obese here.

Conclusion

The anthropometric measurements indicate that mean of height, weight and BMI among male population is 165.58 cms, 54.28 kg and 19.79 respectively; whereas mean of height, weight and BMI among females is 157.33 cms, 46.41 kg and 18.78 respectively which is lesser in comparison to males. Hence, females are shorter and lighter than males. On the other hand, the BMI indicates that females are weaker than males, as most of them are not normal and fall under the category of underweight. 11.37 per cent of the male population falls under the category of CED level whereas this percentage is very high among females. 41.17 per cent of females fall under the CED. When male population is concerned, it is evident that 45.28 per cent of individuals are normal. But female population exhibits a very high distinction from that of male population. Only 17.65 per cent of females are normal. None of the males and females is found obese among the Mawasi.

The findings thus indicate that females of Mawasi tribe are weaker than the males and poor nutrition shows its prevalence more among females. They must be given intensive care to overcome the problems.

References

Awasthi, N, Kumar AP. 1999. Nutritional status of hill primary school children. *Ind J Nutr Diet*, 36: 453-59.

Bailey, KV, and Ferro Luzzi A. 1995. Use of body mass index of adults in assessing individual and community nutritional status. Bull. WHO 73(5):673-680.

Biswas, RK, Kapoor AK 2003. A Study on Mortality Among Saharia – A Primitive Tribe ofMadhya Pradesh. *Anthropologist*, 5(4): 283-290.

Gangadharan, K. 2007. Morbidity and health care in Kerala: A distributional profile and implications, *Indian Journal of Social Development*, Vol.7 (2), pp.195-211.

Geetha, S and Madura Swaminathan (1996), Nutritional status of slum children of Mumbai: A Socio-economic survey, *Economic and Political Weekly*, 6 April 1996, pp. 896-900.

Gopalan, C (1967), Malnutrition in childhood in the tropics, *Britain Medical Journal*, Vol. 4, p. 603.

Jai Prabhakar, S.C. and M.R. Gangadhar. 2009. Nutritional Status of Jenukuruba Tribal Children in Mysore District, Karnataka. *Anthropologist*, 11(2): 83-88.

Jelliff, DB 1966. The Assessment of the Nutritional Status of the Community. WHO Mongr Sec No. 53, Geneva: World Health Organisation.

Khongsdier, R., 2005. BMI and morbidity in relation to body composition: a cross-sectional study of a rural community in North-East India. *British J. Nutr.* Vol. 93: pp. 101-107.

Mallikharjuna Rao, K., N. Balakrishna, N. Arlappa, A. Laxmaiah a G.N.V. Brahmam. 2010. Diet and Nutritional Status of Women in India. J Hum Ecol, 29(3): 165-170.

Naidu, A.N., 1994. Body mass index: A measure of nutritional status in Indian population. *Eur. J. Clin. Nutr.* 48 (suppl. 3): S131-S140.

National Nutrition Monitoring Bureau (2002). Diet and nutritional status of rural population, NNMB technical report No. 21, National Institute of Nutrition, Indian Council of Medical Research, Hyderabad.

NCHS Standards, 1983. Nutrition Monitoring and Assessment. In: T Gopaldas, S Seshadari. Measuring Changes in Nutritional Status. Geneva: World Health Organisation.

Satyanarayan, K., S.S. Rao, G. Radhiah *et al.* 1991. Body mass index and mortality rates. *Nutr.Rev.* (National Institute of Nutrition, Hyderabad), 12(1).

Shakya, SR, Bhandary S, Pokharel PK. 2004. Nutritional Status and Morbidity pattern among governmental primary school children in the East Nepal. *Kathmandu University Medical Journal*, II, 2(4): 307-314.

Shankhala, Aarti, Sankhala A.K., Bhatnagar Bhavna, Singh Alpana. 2004. Impact of Intervention Feeding Trial on Nutritional status of 6-10 years old malnourished children. *Anthropologist*, 6(3): 185-189.

Sharma, Barkha, Mitashree Mitra, Suman Chakrabarty and Premananda, 2006. Nutritional status of preschool children of Raj Gond – A Tribal population in Madhya Pradesh, *Malaysian Journal of Nutrition*, Vol. 12(2), pp. 47-155.

Tewari, Pratiba, Shekhawat Nidhi, Choudhary Sanju. 2005. Use of Nutritional Anthropometry and Clinical Examination in the assessment of Nutritional status of children. *Man in India*, 85(1 and 2): 49-60.

Urade, BP, Moyna Chakravarty, Mallick S.K. 2004. Assessment of Nutritional Status among the Khaire Kunbi children of Maharashtra. *J Hum Ecol*, 15(2): 135-142.

World Health Organisation. 1983. Measuring Change in Nutritional Status. Guidelines for Assessing the Nutritional Impact of Supplementary Feeding Programmes for Vulnerable Groups. Geneva: World Health Organisation.

Joe, William, Mishra U.S., Navaneetham, N. 2009. 'Inequalities in Childhood Malnutrition in India: Some Evidence on Group Disparities.'

Theme – 4
Physical Activity and Body Composition

11

Habitual Physical Activity, Body Composition and Cardio-respiratory Health: Study among Rajput Women of District Kangra, Himachal Pradesh

—*NK Mungreiphy, Meenal Dhall, Satwanti Kapoor*

Introduction

Obesity and physical activity levels are closely interrelated and it is generally accepted that low physical activity levels are associated with increased body weight. The association between physical activity level and obesity has been documented in a number of studies (Di Pietro, 1999, Steinbeck, 2001, Wareham *et al.*, 2005). In 1992, the American Heart Association named physical inactivity as an independent risk factor for cardiovascular disease (Fletcher *et al.*, 1992). Accumulating evidence also indicates that an active and fit way of life helps preserve functional ability and maintain independent living in older adults (Buchner *et al.*, 1992).

In parallel with physical activity, sedentary behaviour is of particular interest, as a complementary dimension related to specific health outcomes (Biddle *et al.*, 2003; Biddle, 2007). Sedentary behaviour refers to occupations that do not increase energy expenditure substantially above the resting level and includes activities such as sleeping, sitting, lying down, watching television and other forms of screen-based entertainment (Pate *et al.*, 2008). The time spent in sedentary behaviour and physical activity levels are generally independently associated with BMI (Biddle, 2007) and other related health outcomes (Helmerhorst *et al.*, 2009; Hamilton *et al.*, 2007).

The existing environments within a country (e.g., the built environment, transport systems, active recreation opportunities, cuisine, food culture and culture around body size) can greatly moderate or modulate the effects of the global obesity drivers on population BMI. These effects can be powerful and help to explain much of the differences in obesity prevalence between populations. They clearly hold opportunities for interventions to make environments less obesogenic (Swinburn *et al.*, 2011). There is an increasing consensus among obesity experts that changing the "obesogenic" environment is a critical step toward reducing obesity. Reversing the factors that lead to increased caloric consumption and reduced physical activity would require major changes in urban planning, transportation, public safety, food production and marketing (Caballero, 2007).

The conventional metabolic coronary heart disease risk factors, such as hypertension, hyperlipidemia, and diabetes, remain powerful predictors of cardiac events in older adults and increase in prevalence and severity with age (Psaty *et al.*, 1999; Tervahauta *et al.*, 1995). Changes in lifestyle and risk factors as a consequence of urban migration might be even greater when migration takes place from less developed rural settings (Torun *et al.*, 2002). Being physically active was related to a decreased risk of CVD mortality, but this association disappeared after controlling for other factors (Kilander, 2001).

Various studies had also been conducted among rural and tribal populations in India. In a study among the Tangkhul Naga tribes, it was found that there has been urbanisation and economic development over the years, a nutritional transition, improved socio-economic status and increasingly sedentary lifestyle, which contributed to the increasing prevalence of overweight/obesity and hypertension among the Tangkhul Nagas (Mungreiphy *et al.*, 2011). Saharia tribe is a socio-economically weaker population with very low literacy level but the clustering of higher blood sugar level, higher blood pressure and higher fat percentage prevalent among them is an indicator of a beginning of metabolic syndrome among this primitive tribal group showing a paradoxical situation (Kapoor *et al.*, 2010). An influence of changing life style among Saharia males was more noticeable with the presence of hypertensive cases among overweight males (Kapoor *et al.*, 2009).

Himachal Pradesh despite its geographical nature has lots of influx of modern life style due to tourism. The females are engaged in their traditional activities as well as jobs i.e. in private and government sectors but the environment remains stressful demanding high level of physical activity. In the light of the global trend of changing lifestyle and its impact on body composition and cardio-metabolic health, there arise needs to assess the morpho-physiological characteristics in relation to their life style characteristics.

Subjects and methods

The present field work was conducted mainly in and around Dharamshala of Kangra District, Himachal Pradesh, which lies in Northern India. Kangra district is situated in Western Himalayas between 31°2 to 32°5 N and 75° to 77°45 E. The field areas include village namely: Naddi, Mcleodganj, Bhagsunag (near Naddi), Khaniara, Chola, Sidhpur, Sidhwari, Yol, Dal, Kunal, Gaggal etc. Cross-sectional study was conducted among 185 females ranging in age from 20-35 years. Informed consent was obtained from each of the subjects. Personal interview was taken and various anthropometric as well as physiological variables were included in the study.

Cardio respiratory functions: Systolic blood pressure, diastolic blood pressure, heart rate, pulse rate, grip strength (right and left hand), random blood sugar, dynamic lung functions (Forced expiratory volume at one second, forced vital capacity, forced expiratory ratio, peak expiratory flow rate), breath holding time, average grip strength, chest expansivity.

Obesity indices: Body mass index, grand mean thickness, waist hip ratio, waist height ratio (Anthropometric measurements i.e. height, weight, waist circumference, hip circumference, chest circumference at normal respiration, maximum inspiration and maximum expiration, skinfold thicknesses at biceps, triceps, subscapular, suprailiac, calf posterior were used to derive cardio respiratory functions and obesity indices).

Standard anthropometry (Tanner, 1969; Martin and Saller, 1957) was used to assess height, body weight, circumferences and skinfolds measurements. All the measurements were taken on the right side of the subjects (Martin and Saller, 1957), except skinfolds, which was taken on the left side of the subjects (Tanner, 1969). Anthropometer was used to measure height. Body weight was taken on each subject with the help of weighing machine. Circumferences were taken with the help of measuring tape. Skinfolds measurements were taken with the help of Holtain skinfold caliper in mm.

In physiological tests, blood pressure (both systolic and diastolic), heart rate, pulse rate, dynamic lung functions, breath holding time and hand grip strength were taken by using standards methods (Consolazio *et al.*, 1963; Weiner and Lourie, 1969). For assessing blood pressure, indirect ausculatory arterial blood pressure of brachial artery was measured with a cuff, a sphygmomanometer and a stethoscope. Heart rate and pulse rate were taken with the help of stopwatch and stethoscope. To assess musculer strength, test was performed on each hand with the help of Dynamometer.

Dynamic lung functions were assessed using portable spirometer by single breath. Forced expiratory volume in first second ($FEV_{1.0}$), forced vital capacity (FVC), forced expiratory ratio (FER), peak expiratory flow rate (PEFR) were taken. $FEV_{1.0}$ defined as the volume of air which can be exhaled in first second of expiratory effort during a forced expiration after a maximum inspiration; FVC is the maximum volume of air that can be exhaled as rapidly as possible from a position of maximum inspiration. Expiratory effort in forced vital capacity is forceful where as in vital capacity the expiratory effort is gentle; FER is the ratio of forced expiratory volume to forced vital capacity and PEFR is peak expiratory flow rate. Breath holding time was counted after putting a nose clip to hold the breath and then time counted with the help of stop watch.

Random blood sugar was measured with single prick using glucometer (Accu check). The care was taken that subject did not had their meals at least two hours of the test. The group with less than 140 mg/dl was taken as having normal blood sugar and more than 140 mg/dl may have diabetes (American Diabetic Association, 2009).

Results

In the study, various social and biological data were taken into account to get a better understanding of the lifestyle and habitual physical activity of the Rajput women and to evaluate its association with their body composition and cardio-respiratory health. Evaluating the type of marriage showed that 87 per cent subjects had arranged marriages while only 13 per cent subjects had love marriages. Regarding family type, 75.7 per cent respondents were in joint families while 24.3 per cent lived in nuclear families. Most (81.1 per cent) of the subjects have their own houses and 18.9 per cent subjects live in rented houses. Concerning food habits, 85.4 per cent subjects were vegetarian while 14.6 per cent subjects had non-vegetarian diet. Most of them (69.7 per cent) had wheat as their staple diet while only 30.3 per cent had rice-based staple diet.

Assessing the type of medication used by the subjects showed that there were 8.6 per cent subjects dependent on traditional medicines whereas 91.4 per cent used modern medication. History of vaccination as known by the subjects was evaluated and 81.6 per cent recalled positively their history of vaccinations but 18.4 per cent were unsure about their vaccinations.

Socio-economic indicators like education, income and occupation were also evaluated. Assessing educational status of the subjects showed that there were 35.1 per cent who were 12th graders and only 6.5 per cent had education below 10th grade. There were 23.8 per cent graduates and 17.8 per cent post graduates among them. Distribution of subjects on the basis of monthly income showed that most of the subjects (34.1 per cent) have monthly income of Rs 5001-10,000 while only 2.2 per cent had a monthly income reaching Rs 20,001 and above. A total of 24.9 per cent earned less than Rs 5000. There were 28.6 per cent and 10.3 per cent subjects whose monthly income categories fall between Rs 10,001-15,000 and Rs 15,001-20,000 respectively. Subject-wise distribution on the basis of occupation shows that 16.2 per cent subjects were in jobs (both private and government) while 83.8 per cent were home-makers who were very active in their traditional activities be it agriculture or weaving etc, besides household chores.

Descriptive statistics of anthropometric measurements has been given in Table 11.1. The mean age of the population under study was 24.8±4.48 years (age range = 20.0-39.0 years), mean height was 155.5±5.70 cm (range = 141.2-168.0 cm) and mean weight was 49.0±8.09 kg (range = 32.5-71.0 kg). The mean waist circumference was 69.9±8.80 cm (range = 54.9-107.0 cm) and the mean hip circumference was 89.1±7.56 cm (range = 68.8-119.0 cm). The mean value for chest circumference at normal respiration was 73.0±8.38 cm while Chest circumference (Maximum inspiration) and Chest circumference (Maximum expiration) was 76.3±8.58 cm and 71.8±8.31 cm respectively.

Skinfolds thicknesses at different sites were also given in Table 11.1. The mean skinfold thickness at sites biceps and triceps was 6.3±3.29 mm and 11.6±4.64 mm respectively. Skinfold thickness at subscapular site and suprailiac site was 13.0±5.77 mm and 14.2±5.66 mm. Mean value for skinfold thickness at calf taken posteriorly was 14.0±5.43 cm.

Table 1. Descriptive statistics for anthropometric measurements

Variables	*Mean ± SD*	*Minimum*	*Maximum*
Age (years)	24.8±4.48	20.0	39.0
Height (cm)	155.5±5.70	141.2	168.0
Weight (kg)	49.0±8.09	32.5	71.0
Waist circumference (cm)	69.9±8.80	54.9	107.0
Hip circumference (cm)	89.1±7.56	68.8	119.0
Chest circumference (Normal respiration)	73.0±8.38	61.0	98.0
Chest circumference at maximum inspiration (cm)	76.3±8.58	63.0	101.0
Chest circumference at maximum expiration (cm)	71.8±8.31	60.0	98.2
Biceps skinfold thickness (mm)	6.3±3.29	2.0	22.0
Triceps skinfold thickness (mm)	11.6±4.64	5.0	24.4
Subscapular skinfold thickness (mm)	13.0±5.77	3.4	45.0
Suprailiac skinfold thickness (mm)	14.2±5.66	4.2	27.8
Calf posterior skinfold thickness (mm)	14.0±5.43	3.0	29.4

Descriptive statistics for physiological measurements is presented in Table 11.1. The mean systolic and diastolic blood pressure was 119.5±11.37 mmHg and 76.0±7.27 mmHg respectively. Mean heart rate and pulse rate of the population were 76.6±7.42 beats/min and 76.3±6.99 beats/min respectively. Random blood sugar varies from 85.0 to 167.0 mg/dl and the mean value was 114.7 ±15.46 mg/dl.

The mean values for hand grip strength were 26.3 ±5.37 kg and 25.3 ±5.21 kg respectively for right and left hand. Right hand grip strength varies from 10.0-41.0 kg and left hand grip strength varied from 12.0-39.0 kg.

Means, standard deviations, minimum and maximum for various lung functions has also been shown in the table. Forced expiratory volume at one second varied from 1.02-3.32 lit./min and forced vital capacity from 1.34-5.61 lit./min. The mean breath holding time was 22.8±9.12 seconds. The means for forced expiratory ratio and peak expiratory flow rate were 87.7±11.10 and 225.6±80.46 liter respectively.

Table 11.2: Descriptive statistics for Physiological measurements

Physiological Measurements	*Mean ± SD*	*Minimum*	*Maximum*
Systolic blood pressure (mmHg)	119.5 ±11.37	90.0	150.0
Diastolic blood pressure (mmHg)	76.0 ±7.27	58.0	96.0
Heart rate (beats/min.)	76.6 ±7.42	56.0	103.0
Pulse rate (pulse/min.)	76.3 ±6.99	55.0	103.0
Right hand grip strength (kg)	26.3 ±5.37	10.0	41.0
Left hand grip strength (kg)	25.3 ±5.21	12.0	39.0
Random blood sugar (mg/dl)	114.7 ±15.46	85.0	167.0
Dynamic lung functions:			
Forced expiratory volume in 1 second (lit./min)	2.30±0.55	1.02	3.32
Forced vital capacity (lit./min)	2.65±0.74	1.34	5.61
Forced expiratory ratio (per cent)	87.7±11.10	57.0	100.0
Peak expiratory flow rate (lit.)	225.6±80.46	73.0	428.0
Breath holding time (seconds)	22.8±9.12	3.21	50.0

Table 11.3: Adiposity indices and muscular strength of the subjects

Derived measurements	*Mean ± SD*	*Minimum*	*Maximum*
Body mass index (kg/m^2)	20.3±2.96	13.2	28.8
Grand mean thickness (mm)	11.8±3.92	4.1	22.2
Waist hip ratio	0.78±0.06	0.67	0.99
Waist height ratio	0.45±0.06	0.35	0.68
Average grip strength (kg)	25.8±5.05	12.0	39.0
Chest Expansivity (cm)	4.4±1.81	0.50	11.7

Table 11.3 shows mean, SD, minimum and maximum of the derived measurements. Body mass index (BMI) is representative of body fatness independent of age, sex, and ethnicity. A measure to

describe the relation between body weight and stature in humans is BMI which is body weight (in kg) divided by stature (in m^2). Body weight adjusted for stature is often used as an alternative to the measurement of adipose tissue mass in the evaluation of individuals or populations for obesity. Various skin-folds have been used to understand the fat distribution pattern and the mean of skinfold thickness was taken as grand mean thickness (GMT). The mean BMI of the subjects was 20.3±2.96 kg/m^2 and grand mean thickness was 11.8±3.92 mm.

The mean value of waist hip ratio was 0.78±0.06 and that of waist height ratio was 0.45±0.06. Waist-to-hip ratio (WHR) and waist to height ratio (WHtR) has also been used as a supplement tool to BMI. WHR and WHtR are increasingly being accepted as one of the best anthropometric indicator of abdominal adiposity and metabolic risk.

Average grip strength is a well established measure of maximum isometric strength of the hand and forearm muscles and is highly correlated to ones muscular strength. Chest expansivity gives the maximum amount of strechability of the chest. The mean value of average grip strength was 25.8±5.05 kg and for chest expansivity it was 4.4±1.81 cm.

Table 11.4: Prevalence of general and regional obesity, and cardio-metabolic health risk

Obesity Indices and Cardio-metabolic health		*N*	*Per cent*
Body mass index	Underweight	56	30.3
	Normal	115	62.2
	Overweight	14	7.6
Waist hip ratio	Risk	60	32.4
	Normal	125	67.6
Waist height ratio	Risk	36	19.5
	Normal	149	80.5
Waist circumference	Risk	27	14.6
	Normal	158	85.4
Systolic blood pressure	Normal	83	44.9
	Prehypertensive	91	49.2
	Hypertensive	11	5.9

(*Contd...*)

Diastolic blood pressure	Normal	121	65.4
	Prehypertensive	53	28.6
	Hypertensive	11	5.9
Random blood sugar	Normal	178	96.2
	Risk	07	3.8

The prevalence of general and regional obesity, and also cardio-metabolic health risk has been given in Table 11.4. Overweight, obesity and thinness were defined using WHO, BMI cut-off values for adults (2004). There were 62.2 per cent subjects with normal BMI, 7.6 per cent were overweight while 30.3 per cent were underweight. According to risk cut offs for central obesity indices, WHR, WHtR and WC, there were 32.4 per cent, 19.5 per cent and 14.6 per cent subjects who fall under risk categories respectively as against 67.6 per cent (WHR), 80.5 per cent (WHtR), and 85.4 per cent (WC) normal subjects.

The Table also shows the distribution of subjects according to systolic and diastolic blood pressure. There were 44.9 per cent and 65.4 per cent subjects with normal blood pressure as assessed from systolic and diastolic blood pressure. As evaluated from systolic blood pressure there were 49.2 per cent and only 5.9 per cent subjects who were prehypertensive and hypertensive respectively. There were 28.6 per cent and 5.9 per cent prehypertensive and hypertensive subjects respectively as assessed from diastolic blood pressure. Distribution of subjects according to random blood sugar level showed 96.2 per cent subjects to have normal random blood sugar level, while only 3.8 per cent subjects were in high category.

Cross-tabulation of BMI with systolic and diastolic blood pressure and blood sugar level is evident from Table 11.5. It is apparent that 13.5 per cent of the underweight subjects were prehypertensive and only 0.5 per cent were hypertensive. Most (16.2 per cent) of the underweight subjects had normal systolic blood pressure whereas most (5.4 per cent) of the overweight were prehypertensive. 27.6 per cent subjects with normal BMI has normal systolic blood pressure and there were 30.3 per cent subjects in prehypertensive category.

Table 11.5: Cross-tabulation of BMI with blood pressure and blood sugar level

BMI		*Systolic blood pressure*				*Diastolic blood pressure*				*Blood sugar level*		
		*I**	*II**	*Normal*	*Total*	*I**	*II**	*Normal*	*Total*	*Risk*	*Normal*	*Total*
Underweight	N	25	1	30	56	13	1	42	56	4	52	56
	per cent	13.50	0.50	16.20	30.30	7.00	0.50	22.70	30.30	2.20	28.10	30.30
Overweight	N	10	2	2	14	6	1	7	14	0	14	14
	per cent	5.40	1.10	1.10	7.60	3.20	0.50	3.80	7.60	0.00	7.60	7.60
Normal	N	56	8	51	115	34	9	72	115	3	112	115
	per cent	30.30	4.30	27.60	62.20	18.40	4.90	38.90	62.20	1.60	60.50	62.20
Total	N	91	11	83	185	53	11	121	185	7	178	185
	per cent	49.20	5.90	44.90	100	28.60	5.90	65.40	100	3.80	96.20	100

I* - Prehypertensive; II* - Hypertensive

Cross-tabulation of BMI with diastolic blood pressure showed that most (22.7 per cent) of the underweight subjects had normal diastolic blood pressure. seven per cent of the underweight subjects were prehypertensive and only 0.5 per cent were hypertensive. Further, 3.2 per cent of the overweight subjects were prehypertensive and 3.8 per cent had normal diastolic blood pressure. A total of 38.9 per cent subjects with normal BMI has normal diastolic blood pressure whereas 28.6 per cent were prehypertensive.

Cross-tabulation of BMI and random blood sugar showed that 28.1 per cent of the underweight subjects had normal blood sugar level while 2.2 per cent were in risk group; whereas 60.5 per cent subjects with normal BMI had normal blood sugar level and only 3.8 per cent were found to have high level of blood sugar.

Table 11.6 presents correlation analyses between various obesity indices and physiological measurements. Body mass index has statistically significant positive correlation with systolic and diastolic blood pressure, breath holding time and FVC. BMI also has significant positive correlation with other obesity indices like GMT, WHR and WHtR. Grand mean thickness has positive correlation only with BMI while with the rest of the obesity indices and physiological measurements there was no significant correlation. Waist hip ratio has statistically significant positive correlation with BMI, WHtR, systolic and diastolic blood pressure. Waist height ratio showed significant positive correlation with BMI, WHR, systolic and diastolic blood pressure.

There were significant positive correlations of systolic and diastolic blood pressure with most of the obesity indices; BMI, WHR and WHtR but do not show significant correlation with most of the physiological measurement including lung functions and blood sugar level. Pulse rate and heart rate do not show significant correlation with most of the obesity indices and physiological measurements except for diastolic blood pressure and blood sugar level. Breath holding time showed significant positive correlation with BMI and average grip strength. Lung functions, FEV_1, FVC, FER and PEF showed significant positive correlations with each other only, while with the rest of the variables no significant correlation was found, except for blood sugar level and FVC. Blood sugar level had significant positive correlations only with

Table 11.6: Correlation matrix of obesity indices and physiological measurements

	BMI	*GMT*	*WHR*	*WHtR*	*SBP*	*DBP*	*Heart rate*	*Pulse rate*	*BHT*	*FEV_1*	*FVC*	*FER*	*PEF*	*Blood sugar*	*AGS*	*CE*
BMI	1	.26**	.42**	.82**	.32**	.18*	-.10	-.12	.21**	.15	.18*	-.04	.06	-.16*	.17*	-.02
GMT	.26**	1	.06	.14	.08	.11	.04	.05	-.02	-.03	-.07	.06	-.05	-.13	.10	.001
WHR	.42**	.06	1	.77**	.21*	.32**	-.04	-.02	.03	.13	.14	.03	.08	-.27**	.08	.02
WHtR	.82**	.14	.77**	1	.36**	.34**	-.06	-.06	.14	.14	.16	.01	.11	-.24**	.14	-.003
SBP	.32**	.08	.21*	.36**	1	.49**	.16	.13	.05	.04	.04	.05	.08	-.07	.06	.13
DBP	.18*	.11	.32**	.34**	.49**	1	.19*	.20*	-.03	.10	.05	.08	.15	.031	.06	.18*
Heart rate	-.10	.04	-.04	-.06	.16	.19*	1	.96**	-.004	.003	.02	-.02	-.07	.17*	-.13	.02
Pulse rate	-.12	.05	-.02	-.06	.13	.20*	.96**	1	.001	-.02	-.02	-.01	-.08	.138	-.11	.01
BHT	.21**	-.02	.03	.14	.05	-.03	-.004	.01	1	-.06	-.06	-.04	.04	-.05	.17*	.01
FEV_1	.15	-.03	.13	.14	.04	.10	.003	-.02	-.06	1	.83**	.27**	.63**	.14	.11	.01
FVC	.18*	-.07	.14	.16	.04	.05	.02	-.02	-.06	.83**	1	-.26**	.29**	.20*	.11	-.05
FER	-.04	.06	.03	.01	.05	.08	-.02	-.01	-.04	.27**	-.26**	1	.61**	-.16*	.03	.08
PEF	.06	-.05	.08	.11	.08	.15	-.07	-.08	.04	.63**	.29**	.61**	1	.01	.11	-.06
Blood sugar	-.16*	-.13	-.27**	-.24**	-.07	.03	.17*	.14	-.05	.14	.20*	-.16*	.01	1	.04	-.05
AGS	.17*	.10	.08	.14	.06	.06	-.13	-.11	.17*	.11	.11	.03	.11	.04	1	.15
CE	-.02	.01	.02	-.003	.13	.18*	.02	.01	.01	.01	-.05	.08	-.06	-.05	.15	1

**p<0.01; *p<0.05 (2-tailed)

heart rate and FVC. Average grip strength showed significant positive correlations with BMI and breathe holding time. Chest expansivity had significant positive correlations only with diastolic blood pressure.

The influence of different occupations on BMI of the subjects was evaluated. Distribution of subjects according to occupation in different BMI categories showed that 53 per cent housewives and only 17 per cent office goers had normal BMI. Home-maker were also usually engaged in their traditional habitual activities. The percentage of those in jobs who were overweight was 5.4 per cent as against 2.2 per cent who were engaged in household work and other traditional habitual activities as housewives. In underweight BMI category there were 28.6 per cent home-makers but only 1.6 per cent constitutes the office goers.

Table 11.7: Regression analysis of obesity indices on occupation

Obesity	*Indices*	*B*	*Sig.*	*Exp(B)*	*95 per cent Confidence Interval for Exp(B)*	
					Lower Bound	*Upper Bound*
BMI	Job	2.67	0.001	**14.41**	4.05	51.25
	Housewife	0^b				
WHR	Job	0.56	0.167	1.76	0.79	3.91
	Housewife	0^b				
WHtR	Job	1.10	0.012	**3.01**	1.28	7.09
	Housewife	0^b				
WC	Job	1.61	0.001	**5.03**	2.03	12.43
	Housewife	0^b				

a. The reference category is: normal.

b. This parameter is set to zero because it is redundant.

Table 11.7 shows regression analyses for obesity indices which include BMI, WHR, WHtR and WC with occupation. The regression of BMI with occupation showed that those who were engaged in office jobs besides household chores were associated with 14.41 times higher chances of being overweight as against those involved in traditional habitual activities besides household chores.

The Table also shows regression coefficients for waist hip ratio with occupation. There are no clear significant associations between waist hip ratio and occupation either having a job or being a housewife. Regression analyses for waist height ratio with occupation show that being in office job is associated with 3.01 times higher risk of central obesity than those who were housewives. As assessing from regression coefficients for waist circumference with occupation, office goers had 5.03 times higher risk of becoming centrally obese than housewife.

Discussion

In present study, the relationship between body composition, cardio-respiratory health and habitual physical activity was assessed among Rajput women of District Kangra, Himachal Pradesh. Most of the subjects were homemakers but were physically very active in their traditional ways of livelihood, be it agriculture or weaving. The study showed that most of the subjects had normal BMI, followed by underweight but there were very few overweight subjects also. Assessment from central obesity indices also showed that most of the subjects were in normal category. Blood pressure as assessed from systolic showed that most of them were prehypertensive but when assessed from diastolic blood pressure most of the subjects were in normal category. Most of the subjects had normal random blood sugar level.

Most of the Rajput women had normal body mass index which corresponds to their good cardio-respiratory and metabolic health as assessed from systolic and diastolic blood pressure, lung functions and random blood sugar level. These women were home-makers and besides household chores they were very active in their traditional activities, the habitual physical activity was evident by their prolonged working hours involving both outdoor and household chores. Their physically active lifestyle worked in their favour in preventing lifestyle diseases, which has been prevalent among transitional and urban populations. The accumulation of these physical activity in association with lifestyle physical activity has already been recommended as the exercise treatment for hypertension (Pescatello *et al.*, 2004) as well as for the prevention of prehypertension progressing to hypertension (Chobanian *et al.*, 2003). Besson *et al.*(2009) showed a strong inverse relationship between total physical activity and body mass index (BMI), independent of age, gender and country, as well as other potential confounding factors, including energy intake. Physical activity is recognised as a complex behaviour, with the main characteristics being type, duration, frequency and intensity.

Rajputs of Himachal Pradesh are isolated from main stream of urban population earlier but as they try to adapt and interact with urban population, it may affect their lifestyle. Besides this, majority of them worked in fields or were engaged in habitual physical activities for 4-6 hours a day with a physical activity level (PAL)>1.4 (James *et al.*, 1988). Other communities like Rajis of Uttarakhand and Saharias of Madhya Pradesh also had the same pattern of PAL (Kapoor *et al.*, 2008; Kapoor *et al.*, 2009). Among Rajis and Saharia, they were mostly CED (Chronic Energy Deficient) but had better muscular strength, while Rajputs women mostly had normal BMI with muscularity. This may imply that Rajputs females of HP, highlighting the adaptive role of higher level of habitual physical activity and better nutritional status enjoyed the same cardio-respiratory health as Rajis; whereas Saharias showed onset of metabolic disorder and hypertension.

Before the twentieth century, the traditional diet of these populations composed almost exclusively on plant foods and animal products when available but this diet has become progressively westernised, with an increase in the intake of carbohydrates and fat. The exchange of nutritional attributes with surrounding populations has lead to such a trend. Only the communities which have accepted the idea of this nutritional substitution have now been associated with a certain degree of lifestyle disorders and increase in percentage of populations falling under risk categories of obesity markers. Similar results have been shown in the present study. Globalisation has put forth numerous opportunities and jobs. This occupational change has also fuelled nutritional dynamics putting these individuals at a higher risk of being overweight unlike traditional workers (home-makers) which show no significant association with being overweight and/or having high blood sugar levels.

Along with undernutrition, there coexist the problem of overweight/obesity and associated health problems prevalent among tribal groups like Tangkhul Nagas and Saharia tribes which have started coming in touch with the mainstream populations, adopting their way of life (Mungreiphy *et al.*, 2011; Kapoor *et al.*, 2010). Whereas more isolated tribe group like Rajis mostly had lean BMI so the problems of overweight/obesity and associated health problem was not noticed in them. Moreover, Rajis mainly are habitually very active (Kapoor *et al.*, 2009).

Kapoor *et al.*, (2010) showed how Saharias, a tribal group in Madhya Pradesh display high blood sugar levels. They often migrate to nearing towns for work; get exposed to modern means of life and food items which might partially explain the present scenario. On the other hand, Raji, primarily a food gatherer and hunter tribe of Indian Himalayas have still maintained their distinct cultural and occupational identity due to geographical isolation and have no apparent change in their dietary or physical activity pattern and are still free from the clinches of falling in risk categories of obesity-markers or other lifestyle diseases.

Most lifestyle disorders are a complex matrix that emerge from changes in diet (Maddocks, 1961), changing ethical, social, and moral values (Henry and Cassel, 1963), crowding (Scotch and Geiger, 1963) anxiety engendered by entry into a differently oriented society, economic stress, competition, racial and ethnic tensions, noise, pollution, pace of life, etc. (Scotch and Geiger, 1969). In most reported studies, the multiplicity of simultaneous forces acting on these communities makes it difficult to assess the importance of any individual influence. People involved in the traditional occupation tend to have a traditional diet, characterised by a relatively higher intake of vitamins and minerals as compared to those living with modern lifestyle which in turn is directly associated with predisposition to lifestyle disorders as well (Ross *et al.*, 2006). This certainly does not mean that one revert backs to the older lifestyle but it puts forward a very pertinent point -whether we in our present modern occupations along with the voluntary physical activity levels do match the energy requirement equivalent to those populations still engaged in their tradition high level energy expenditure activities? The present population does prove this point of contention that despite being exposed to the populations from plains, experiencing exposure to fast food and lifestyle change, majority retains their traditional level of energy expenditure, the effect of which is reflected in their nutritional status and cardio-respiratory and metabolic health.

It is also known that community average blood pressures in more developed, industrialised societies are higher than those in less developed, traditional societies. These differences can be accounted for only in part by differences in physical activity, body composition and nutrient intake. A portion of the difference is accounted for by more direct effects of culture on individual behaviour and physiology (Dressler, 2004). The nutrition intake of traditional societies is changing and so is their body composition. What has remained unaffected is their physical activity level and hence, their health parameters have remained unfazed. Human cultural development has often been divided into three phases by anthropologists, the hunter-gatherers, the traditional workers and modern industrialised world. The former categories have always had very high levels of energy expenditure and hence forth obesity is rare (Bouchard and Katzmarzyk, 2010). The additive affects of occupational physical activity with traditional chores have nullified the nutritional dynamics that have occurred due to global exposure.

Acknowledgements

The financial assistance by UGC-SAP is greatly acknowledged, under which this project was undertaken. The cooperation of all the subjects and key informant Ms Neelam, and fields guides Mr Sonu, Mr Suresh and Mr Pawan are genuinely acknowledged. The inputs from Dr Mary Grace Tungdim, Ms Deepali Verma, Ms Heemansu Aurora, Ms Savita Goswami and Mr Vineet Kumar Verma are deeply acknowledged.

References

Andersen, RE, Wadden TA, Bartlett SJ, Zemel B, Verde TJ, Franckowiak SC. Effects of Lifestyle Activity vs Structured Aerobic Exercise in Obese Women. JAMA. 1999; 281:335-340.

Aubert, L, Bovet P, Gervasoni JP, *et al.*, 1998. Knowledge attitudes and practices on hypertension in a country in epidemiological transition. *Hypertension* 31:1136- 1145.

Besson, H, Ekelund U, Luan J, May AM, Sharp S, Travier N *et al.*, A cross-sectional analysis of physical activity and obesity indicators in European participants of the EPIC-PANACEA study. *Int J Obes* (Lond) 2009; 33: 497-506.

Biddle, S, Gorely T, Marshall SJ, Murdey I, Cameron N. Physical activity and sedentary behaviours in youth: issues and controversies. *J Roy Soc Promot Health* 2003; 124: 29-33.

Biddle, SJ. Sedentary behavior. *Am J Prev Med* 2007; 33: 502-504.

Blair, SN, Kohl HW III, Gordon NF. Physical activity and health: a lifestyle approach. *Med Exerc Nutr Health*. 1992;1:54-57.

Bouchard, C, Perusse L, LeBlanc C, Tremblay A, Theirault G: Inheritance of the amount and distribution of human body fat. *M J Obes* 1988;12: 205-215.

Bouchard, C, Tremblay A, Despres JP, Nodeau A, Lupien PJ, Theriault G, Dussault J, Noorjani S, Pinault S, Fournier G. The response to long-term overfeeding in identical twins. *N Engl J Med* 1990; 322:1477-1482.

Bouchard, C. and Katzmarzyk PT. Physical activity and obesity. 2nd ed. Baton Rouge, LA: Human Kinetics; 2010:210-213.

Brownell, KD, WaddenTA. Etiology and treatment of obesity. *J Consult Clin Psychol*. 1992; 60:505-517.

Buchner, DM, Beresford SAA, Larson EB, LaCroix AZ, Wagner EH. Effects of physical activity on health status in older adults, II: intervention studies. *Annu Rev Public Health*. 1992;13:469-488.

Caballero, B. A nutrition paradox: underweight and obesity in developing countries. *N Engl J Med* 2005; 352:1514-16.

Caballero, B. The Global Epidemic of Obesity: An Overview. *Epidemiol Rev* 2007; 29:1-5.

Census of India 2001 and 2011; www.censusindia.gov.in/

Cesari, M, Penninx BW, Newman AB, *et al.*, Inflammatory markers and onset of cardiovascular events: results from the Health ABC study. *Circulation* 2003;108:2317-22.

Chobanian, A. V., Bakris G. L., Black H. R., *et al.*, JNC7- Complete Report: Seventh report of the Joint National Committee on prevention, detection, evaluation, and treatment of high blood pressure. *Hypertension* 42:1206-1252, 2003.

Church, TS, Thomas DM, Tudor-Locke C, Katzmarzyk PT, Earnest CP, *et al.* (2011). Trends over 5 Decades in U.S. Occupation-Related Physical Activity and Their Associations with Obesity. *PLoS ONE* 6(5): e19657. doi: 10.1371/journal.pone. 0019657.

Cohen, AM, Vably S, Poznanski R. Changes in diet of Yemenite Jews in relation to diabetes and ischaemic heart disease. *Lancet* 1961; 2:1399-1401.

Cushman M, Arnold AM, Psaty BM, *et al.*, C-reactive protein and the 10-year incidence of coronary heart disease in older men and women: the cardiovascular health study. *Circulation* 2005;112:25-31.

DiPietro, L. Physical activity in the prevention of obesity: current evidence and research issues. *Med Sci Sports Exerc* 1999; 31 (11 Suppl): S542–S546.

Dressler, W.W Culture and the risk of disease. *Br Med Bull* (2004) 69 (1): 21-31.

Eckel, RH, Grundy SM, Zimmet PZ. The metabolic syndrome. *Lancet* 2005; 365:1415-28.

Fletcher, GF, Blair SN, Blumenthal J, *et al.*, Position statement: statement on exercise: benefits and recommendations for physical activity programs for all Americans. *Circulation*. 1992; 86:340-344.

Goldstein, DJ. Beneficial health effects of modest weight loss. *Int J Obes Relat Metab Disord* 1992;16:397-415.

Goran, MI, Weinsier RL. Role of environmental vs. metabolic factors in the etiology of obesity: time to focus on the environment. *Obes Res* 2000; 8:407-9.

Guo, SS, Zeller C, Chumlea WC, and Siervogel RM. Aging, body composition, and lifestyle: the Fels Longitudinal. *Am J Clin Nutr* 1999; 70:405-11.

Hamilton, MT, Hamilton DG, Zderic TW. Role of low energy expenditure and sitting in obesity, metabolic syndrome, type 2 diabetes, and cardiovascular disease. *Diabetes* 2007; 56: 2655-2667.

Harding, S. Mortality of migrants from the Indian subcontinent to England and Wales: effect of duration of residence. *Epidemiology* 2003;14:287-92.

Harris, MI. Prevalence of non-insulin-dependent diabetes and impaired glucose tolerance. In: Diabetes in America. U. S. Department of Health and Human Services. NIH Publication No. 85-1468, VI 1-31, Washington D. C. 1985.

Harris, TB, Ballard-Barbasch R, Madan J, Makuc DM, Feldman JJ. Overweight, weight loss, and risk of coronary heart diseases in older women. The NHANES I Follow-up study. *Am. J. Epidemiol.* 1993; 137: 1318 1327.

Helmerhorst, HJ, Wijndaele K, Brage S, Wareham NJ, Ekelund U. Objectively measured sedentary time may predict insulin resistance, independent of moderate and vigorous physical activity. *Diabetes* 2009; 58: 1776-1779.

Henry, JP, Cassel JC: Psychosocial factors in essential hypertension. *Am J Epidemiol* 90: 171, 1969.

Himachal Official Webportal, Himachal Pradesh Government, India. www.himachal.nic.in/ Retrieved on 26-3-2012.

Hu, FB, Li TY, Colditz GA, Willett WC, Manson JE. Television watching and other sedentary behaviors in relation to risk of obesity and type 2 diabetes mellitus in women. *JAMA* 2003; 289: 1785-1791.

James, WPT, Mascie-Taylor CGN, Norgan NG, Bristrian BR, Shetty P, Ferro-Luzzi A. The value of arm circumference measurements in assessing chronic energy deficiency in Third World adults. *Eur J Clin Nutr* 1994; 48: 883-4.

Kangra Official website, http://hpkangra.nic.in/ Retrieved on 26-3-2012.

Kapoor, A.K., Tyagi R, Kapoor S. Nutritional status and cardio-respiratory functions among adult Raji males, a hunter and gatherer tribe of the Indian Himalayas. *Anthropological Science*, Vol. 117(1), 1–7, 2009.

Kapoor, S, Tyagi R, Saluja K, Chaturvedi A, Kapoor A.K. Emerging health threats among a primitive tribal group of Central India. *Journal of Public Health and Epidemiology* 2010; 2(2):13-19.

Kapoor, S, Tyagi R, Saluja K, Chaturvedi A, Kapoor A.K. Nutritional Profile and Socio-Economic Status of Saharia, a Primitive Tribe of India. *The Open Anthropology Journal*, 2008, 2, 58-63.

Kasl, SV, Berkman L. Health consequences of the experience of migration. Annu Rev Public Health 1983; 4:69-90.

Kilander, L, Berglund L, Boberg M, Vessby B and Lithell H. Education, lifestyle factors and mortality from cardiovascular disease and cancer. A 25-year follow-up of Swedish 50-year-old men. *International Journal of Epidemiology* 2001; 30:1119-1126.

Klein, S, Burke LE, Bray GA, *et al.*, Clinical implications of obesity with specific focus on cardiovascular disease: a statement for professionals from the American Heart Association Council on Nutrition, Physical Activity, and Metabolism: endorsed by the American College of Cardiology Foundation. *Circulation* 2004;110:2952-67.

Knowler, WC, Barrett-Connor E, Fowler SE, *et al.*, For the Diabetes Prevention Program Research Group. Reduction in the incidence of type 2 diabetes with lifestyle intervention or metformin. *N Engl J Med*. 2002; 346: 393-403.

Lopez, AD. Assessing the burden of mortality from cardiovascular disease. World *Health Stat Q* 1993; 46:91-96.

Lucia, P, Edoardo C, Carvalho BLM, Mikolaj W, Massimo P, Paolo P, Achille CP. Effects of a traditional lifestyle on the cardiovascular risk profile: the Amondava population of the Brazilian Amazon. Comparison with matched African, Italian and Polish populations. *Journal of Hypertension* 1999; 17 (6): 749-756.

Luke, A, Durazo-Arvizu R, Rotimi C, Prewitt T, Forrester T. Relation between body mass index and body fat in black population samples from Nigeria, Jamaica and the United States. *Am J Epidemiol* 1997; 67:391-96.

Maddocks, I: Possible absence of essential hypertension in two complete pacific island populations. *Lancet* 2: 396, 1961.

Martin, R, and Saller, K. Lehrbuch der Anthropologie. Stuttgart: Gustav Fishera. 1957.

McKeigue, PM. Metabolic consequences of obesity and body fat pattern: lessons from migrant studies. *Ciba Found Symp* 1996; 201:54-64.

Mendez, MA, Monteiro CA, Popkin BM. Overweight exceeds underweight among women in most developing countries. *Am J Clin Nutr* 2005; 81:714-21.

Mungreiphy, N. K., Kapoor S, Sinha R. Association between BMI, blood pressure and age: Study among Tangkhul Naga Tribal males of Northeast India. *Journal of Anthropology*, 2011; ID 748147: 1-6.

Murray, CJL, Lopez AD. Global Comparative Assessments in the Health Sector. Geneva, Switzerland: World Health Organisation, 1994.

NCEP - Executive Summary of the third Report of the National Cholesterol Education Program (NCEP) Expert Panel on Detection, Evaluation, and Treatment of High Blood Cholesterol In Adults (Adult Treatment Panel III). *JAMA* 2001; 285:2486-97.

Pate, RR, O'Neill JR, Lobelo F. The evolving definition of 'sedentary'. *Exerc Sport Sci Rev* 2008; 36: 173-178.

Pate, RR, Pratt M, Blair SN, *et al.*, Physical activity and public health: a recommendation from the Centers for Disease Control and Prevention and the American College of Sports Medicine. *JAMA*. 1995; 273: 402-407.

Pescatello, L. S., Franklin B. A., Fagard R., Farquhar W. B., Kelley G. A., and RAY C. A. American College of Sports Medicine position stand: exercise and hypertension. *Med. Sci. Sports Exerc;* 2004; 36:533–553.

Popkin, BM, Gordon-Larsen P. The nutrition transition: worldwide obesity dynamics and their determinants. *Int J Obes Relat Metab Disord* 2004; 28(suppl):S2-9.

Popkin, BM. The nutrition transition: an overview of world patterns of change. *Nutr Rev* 2004; 62:S140-3.

Prentice, AM, Jebb SA. Obesity in Britain: gluttony or sloth? *BMJ*. 1995; 311: 437-439.

Price, RA, Charles MA, Pettitt DJ, Knowler WC. Obesity in Pima Indians: large increases among post World War II birth cohort. Am Phys Anthmpol 1993; 92:473-479.

Psaty, BM, Furberg CD, Kuller LH, *et al.*, Traditional risk factors and subclinical disease measures as predictors of first myocardial infarction in older adults: the Cardiovascular Health Study. *Arch Intern Med* 1999; 159:1339-47.

Ravussin, E, Valencia M E., Esparza J, Bennett P H., Schulz Leslie O. Effects of Traditional Lifestyle on Obesity in Pima Indians. *Diabetes Care* 1994; 17:9.

Reddy, KS, Yusuf S. Emerging epidemic of cardiovascular disease in developing countries. *Circulation* 1998; 97:596-601.

Ross, Johansson A, Ingman M and Gyllensten U. Lifestyle, Genetics, and Disease in Sami Alastair *B Croat Med J*. 2006 August; 47(4): 553-565.

Scotch, NA, Geiger HJ. The epidemiology of essential hypertension: A review with special attention to psychological and sociocultural factors II psychologic and sociocultural factors in etiology. J Chron Dis 16: 1183, 1963.

Seidell, JC. Time trends in obesity: an epidemiological perspective. *Horm Metab Res*. 1997; 29:155-158.

Sharma, S, Cade J, Landman J, Cruickshank JK. Assessing the diet of the British African-Caribbean population: frequency of consumption of foods and food portion sizes. *Int J Food Sci Nutr* 2002; 53:439-44.

Shear, CL, Freedman DS, Burke GL, Harsha DW, Webber LS, Berenson GS: Secular trends of obesity in early life: the Bogalusa Heart Study. *Am Public Health* 1988; 78: 75-77.

Steinbeck, KS. The importance of physical activity in the prevention of overweight and obesity in childhood: a review and an opinion. *Obes Rev* 2001; 2: 117-130.

Stevens, J, Evenson KR, Thomas O, *et al.*, Associations of fitness and fatness with mortality in Russian and American men in the lipids research clinics study. *Int J Obes Relat Metab Disord* 2004; 28:1463-70.

Stunkard, AJ, Harris JR, Pederson NL, Mc-Clearn GE. The body mass index of twins who have been reared apart. N Engl J Mcd 1990; 322:1483-1487.

Swinburn, BA, Sacks G, Hall KD, McPherson K, Finegood DT, Moodie ML, Gortmaker SL. Obesity 1: The global obesity pandemic: shaped by global drivers and local environments. *Lancet* 2011; 378: 804-14.

Taylor, R, Badcock J, King H. Dietary intake, exercise, obesity and noncommunicable disease in rural and urban populations of three Pacific Island countries. *J Am Coll Nutr* 1992;11:283-93.

Taylor, R, Bennett P, Uili R, Joffres N, Levy S, Zimmet P: Diabetes in Wallis Polynesians: a comparison of residents of Wallis Island and first generation migrants to Noumea, New Caledonia. *Diabetes Res Clin Pract* 1985; 1: 169-178.

Tervahauta, M, Pekkanen J, Nissinen A. Risk factors of coronary heart disease and total mortality among elderly men with and without preexisting coronary heart disease. Finnish cohorts of the Seven Countries Study. *J Am Coll Cardiol* 1995; 26:1623-9.

Torun, B, Stein AD, Schroeder D, Grajeda R, Conlisk A, Rodriguez M, Mendeza H and Martorellb R. Rural-to-urban migration and cardiovascular disease risk factors in young Guatemalan adults. *International Journal of Epidemiology* 2002; 31:218-226.

US Department of Health and Human Services. Physical Activity and Health: A Report of the Surgeon General. Atlanta, Ga: US Dept of Health and Human Services, Centers for Disease Control and Prevention, National Center for Chronic Disease Prevention and Health Promotion; 1996.

Wareham, NJ, van Sluijs EM, Ekelund U. Physical activity and obesity prevention: a review of the current evidence. *Proc Nutr Soc* 2005; 64: 229-247.

Weng, X, Liu Y, Ma J, *et al.*, Use of body mass index to identify obesity-related metabolic disorders in the Chinese population. *Eur J Clin Nutr* 2006; 60: 931-7.

World Health Organisation. Obesity: preventing and managing the global epidemic. Report of a WHO Consultation. Geneva, Switzerland: World Health Organisation, 2000. (WHO technical report series 894).

World Health Organisation/Food and Agriculture Organisation of the United Nations. Diet, nutrition and the prevention of chronic diseases. Report of a Joint WHO/FAO Consultation. Geneva, Switzerland: World Health Organisation, 2003. (WHO technical report series 916).

World Health Organisation: Diet, Nutrition and the Prevention of Chronic Diseases. Report of WHO Study Group, Series 797. Geneva, World Health Organisation, 1990.

12

A Study of Physical Activity Level Among the Chamba Youth (17-18 Years) of Himachal Pradesh

—*Lakhwinder Kaur and Ramesh Sahani*

Abstract

Physical activity level (PAL) is one of the important determinants of health status among the adults and youth. It is inversely correlated with several metabolic disorders, hence, recommended for preventing the rapid increasing level of diseases. It is influenced by several factors. The recent trends of declining PAL may be one of the contributing factors of such rapid increase of metabolic diseases. There is no dearth of scientific studies focusing on PAL and its negative and positive impacts, but population-based studies are still limited. Numerous studies are also available from our country on these aspects, but there is dearth of information from Himacal Pradesh. Hence, present study is conducted among the youth of Chamba, age ranging from 17-18 years, both boys and girls. The study is of cross-sectional nature, information related to different activities performed by them for two consecutive days were noted with the time duration. The data were analysed using SPSS software version 18 to get the descriptive statistics and to see the bisexual differences. The physical activity ratio is computed from it and PAL was derived according to FAO/ WHO/UNU (2001). The result shows both the boys and girls are under the category of very high PAL, girls have significantly higher mean PAL than the boys, contradicting most of the studies. It can be concluded that Chamba youth are leading a very active life-style, also reflected by the low level of obesity.

Key words: *Physical activity level, metabolic disorders, Chamba, blood pressure, obesity.*

Introduction

The second half of the twentieth century witnessed major health transitions in our country, propelled by socio-economic and technological changes that profoundly altered life-expectancy and ways of living. Obesity, cardiovascular diseases (CVD), diabetes mellitus, bone and joint diseases are growing contributors to disease burdens worldwide and India is also experiencing the same condition. Our country also experiencing a rapid change in economy, lifestyle, and the emerging consequences are the rapid increasing of health burden, alongwith other social and mental issues. Besides, other contributing factors, physical activity is also one of the most important contributors for the rapid increase of such metabolic disorders. Physical activities not only helps us to control our blood pressure but also to manage our weight, stress level and also strengthen our heart by making it free from high blood pressure (Jansen and LeBlanc, 2010). It is believed that a healthy weight, strong heart and emotional health are all good for our heart. A healthy weight with the BMI (body mass index) should be or around <23 kg/m^2 and low waist stature ratio (should be or around <50) is considered as normal, can be easily achieved by regular 30 minutes moderate physical activity. There is a negative correlation of blood pressure with 30 minutes moderate physical activity i.e. if the time duration decreases for physical activity, the chances to get high blood pressure increases and if the time duration increases for physical activity, the chances to get high blood pressure decreases. Even, different types of cancer can be prevented adopting the high physical activity level, besides keeping body tenacity well and good.

Numerous studies are available focusing the PAL alone or its correlation with health and diseases, and its positive effects world over, but population-based study is still lacking. It is the well known fact that the PAL is gradually decreasing, even, among the traditional and rural societies. The cause of this may be several. The PAL gain recent attention in our country with the rapid increase of metabolic disorders, but comparatively studies in this aspect is lacking from the rural and traditional societies. The Indian scenario shows worse and very alarming situation in the context of PAL, only 10 per cent participnts is reported in the recreational physical activity (Anjana *et al.*, 2014). The low level of physical activity is also reported from Kerala, the highly educated state of India (Aslesh, 2015), and other studies also showing the same trend (Rastogi, 2004). Swaminathan *et al.*, 2011 concluded in a study "there was a significant decline in moderate-to-vigorous physical activity in urban school children largely due to a decrease in physical activity at school. This was particularly true of girls and older children. The study by Vaz and Bharathi, 2004, suggest clear gender and age determinants of the perceived intensity of specific activities, which determine the physical activity level. The study on PAL from Himachal Pradesh is almost lacking.

Considering the above discussed information and the rapid increase of metabolic syndromes, even, among the youth and children of our country, the present study is conducted among the college students of Chamba, Himachal Pradesh, both males and females to assess the physical activity level and sex difference in it.

Material and Method

Area and people

Himachal Pradesh is a hill state in the north-west parts of India and is famous for its abundant natural beauty. The literal meaning of the Himachal Pradesh is "Land of Snowy Mountains" and is also known as "Dev Bhoomi"- The Land of Gods. It is spread over an area of 55,780 sq. km. and situated between 30° 22' and 30° 12′ north latitude and between 75° 47′ and 79° 4′ east longitude. The economy of the state is highly dependent on three sources: hydroelectric power, tourism and horticulture. The major communities of the state are *Brahmin, Rajput, Kanets, Rnathis and Kolis.* Chamba is the northwestern district of Himachal Pradesh, with its headquarters in Chamba town. It is situated between north latitude 32° 11′ 30" and 33° 13′ 6" and east longitude 75°49 and 77° 3′ 30", with an estimated area of 6528 sq. km. and is surrounded on all sides by lofty hill ranges. The territory is wholly mountainous with altitude ranging from 2,000 to 21,000 feet. The temperatures in summer vary between 38°C and 15°C and in winter: 15°C and 0°C. The average annual rainfall in the town is 785.84 mm. Chamba District is known for its various temples as well as some historic sites. It is one of the popular tourist sites of the state. The staple food is *wheat, maize, barley* which is local produce and the principal pulses consumed are *peas* and *rajmahs.*

The field work was conducted from 17th to 15th September 2014 at government postgraduate college, Chamba among across caste group. To estimate the physical activity level, cross-sectional study is done based on a sample of college students, 100 girls and 101 boys ranging in age from 17 to 18 years. Purposive sampling was used for selecting students of various ages. Students were asked to give the time duration for each physical activity they performed in the last two days. Recall method was applied to get information about the activities performed and time duration of the activities. The activities performed by the subjects were noted with its time duration. The energy cost of each activities i.e. energy spent on the activities, the physical activity ratio (PAR) using the table of FAO/ WHO/UNU (2001) is computed by multiplying the time spent. The PAL is calculated by dividing the total physical activity ratio (PAR) by the whole period of the days in hours i.e. 24 hours and classification of PAL is done (FAO/ WHO/UNU, 2001).

Result

The Table 12.1 gives the details of the activities performed by the Chamba youth. The time spent on sleeping is the highest and girls spent significantly more time for sleeping than the boys. The time spent for sleeping is 8.20 hours and 7.12 hours for girls and boys respectively. The physical activity ratio (PAR) does not differ significantly but higher value is noticed among the girls for sleeping (Table 12.2). The second activity which is followed by sleeping is the sitting and time spent does not differ according to sex for this activity. On an average girls spend 4.7 hours for sitting while in case of boys, the value is 4.5 hours. The PAR value of sitting does not differ significantly between the sexes. Walking is also an important activity among the studied group. The time spent for attending school can also be included in these activities, since most of the youths are coming to the college by walking only. The time spent for this activity

Table- 12.1: Details of time spent for different activities

Activities	Girls						Boys					
	N	Min.	Max.	Mean	SD	SE	N	Min.	Max.	Mean	SD	SE
Sleeping	100	6.00	10.00	8.20	0.91	0.09	101	6.00	10.00	7.72	0.84	0.08
Personal care	100	0.25	2.00	1.08*	0.43	0.04	101	0.25	2.00	0.79	0.33	0.03
Eating	100	0.25	3.00	0.81*	0.42	0.04	101	0.25	2.00	1.14	0.43	0.04
Cooking	92	0.25	5.00	1.51	0.90	0.09	93	0.50	4.00	1.46	0.73	0.08
Sitting	100	1.00	8.00	4.70	1.41	0.14	101	1.00	7.00	4.50	0.98	0.10
General household work	86	0.25	5.00	1.56	1.00	0.11	100	0.25	4.00	1.34	0.65	0.06
Walking	89	0.25	6.00	1.91*	1.48	0.16	101	1.00	6.00	2.35	1.01	0.10
Light leisure activities	97	0.50	6.00	2.35*	1.14	0.12	101	1.00	5.00	1.85	0.85	0.08
Standing carrying light weight	94	0.25	6.00	1.44	1.11	0.11	101	0.50	4.00	1.32	0.57	0.06
Reaching schools (walk)	100	0.25	3.00	1.00*	0.62	0.06	100	0.25	3.00	1.23	0.77	0.08
low intensity arobic exercise	41	0.25	2.00	0.79*	0.48	0.08	100	0.25	3.00	1.02	0.65	0.07
Planting, weeding etc	36	0.25	2.25	0.66	0.55	0.09	13	0.25	2.00	0.75	0.53	0.15
Collecting water/wood	35	0.25	1.50	0.62	0.35	0.06	31	0.25	3.00	0.76	0.64	0.11
Non mechanised domestic works	87	0.50	3.00	1.66*	0.77	0.08	64	0.25	2.00	0.71	0.44	0.06
shopping	100	0.00	5.00	1.78*	1.27	0.13	101	0.00	3.00	0.82	0.62	0.06
Reading	99	1.00	6.00	3.79	1.21	0.12	101	1.00	7.00	3.67	1.30	0.13
Watching TV	97	0.25	5.00	2.78*	0.96	0.10	99	0.50	6.00	2.45	1.08	0.11
Running small distance	69	0.25	2.00	0.71	0.46	0.06	47	0.25	3.00	0.55	0.50	0.07
Swimming	100	0.00	2.00	0.10	0.32	0.03	101	0.00	2.00	0.09	0.29	0.03
Dancing	54	0.25	3.00	1.00*	0.72	0.10	16	0.25	2.00	0.69	0.44	0.11
Climbing stairs	100	0.00	2.00	0.45	0.41	0.04	101	0.00	1.50	0.41	0.25	0.02
Listening Radio/music	90	0.25	3.00	1.79	0.82	0.09	93	0.50	3.50	1.75	0.72	0.07
Paintings	26	0.25	2.00	0.99	0.57	0.11	15	0.50	2.50	0.82	0.62	0.16
Loading of goods on vehicle	100	0.00	3.00	0.09*	0.39	0.04	101	0.00	3.00	1.00	0.69	0.07

(Contd...)

Ironing clothes	90	0.25	1.25	0.71*	0.32	0.03	73	0.25	1.00	0.43	0.21	0.02
Grooming animals	100	0.00	1.50	0.12	0.26	0.03	101	0.00	2.00	0.16	0.40	0.04
Running small distance	32	0.25	2.00	0.70	0.51	0.09	23	0.25	2.50	0.88	0.59	0.12
Sitting	100	0.00	6.00	1.81*	1.53	0.15	101	1.00	10.00	4.32	1.62	0.16
Gardening	31	0.25	1.00	0.60	0.25	0.04	12	0.25	1.00	0.69	0.34	0.10
Washing hands and face	99	0.25	3.00	0.59*	0.39	0.04	100	0.25	1.00	0.43	0.20	0.02
Other domestic activities	99	0.25	4.00	2.24	0.88	0.09	101	0.50	6.00	2.23	1.02	0.10
Talking	100	0.25	5.00	2.09	0.98	0.10	101	0.50	5.00	2.15	1.08	0.11
Lying on bed	100	0.50	4.00	2.40	0.75	0.08	101	0.50	4.00	2.39	0.88	0.09

* Mean differ significantly between boys and girls.

Table 2. PAR values of different activities of the Chamba youth

Activities	*Girls*						*Boys*					
	N	*Min.*	*Max.*	*Mean*	*SD*	*SE*	*N*	*Min.*	*Max.*	*Mean*	*SD*	*SE*
Sleeping	100	6.00	81.00	9.48	0.91	9.11	101	6.00	64.00	8.76	0.74	7.39
Personal care	100	0.70	5.60	3.00*	0.12	1.22	101	0.70	5.60	2.19	0.09	0.93
Eating	100	0.40	4.80	1.26*	0.06	0.64	101	0.40	3.20	1.80	0.07	0.68
Cooking	100	0.00	10.80	3.05	0.21	2.14	101	0.00	8.64	2.86	0.18	1.77
Sitting	100	1.20	24.00	5.87	0.24	2.45	101	1.20	42.00	6.04	0.47	4.72
Gen. Household work	100	0.00	14.00	3.55	0.29	2.88	101	0.00	11.20	3.75	0.19	1.87
Walking at various places without a load	100	0.00	27.00	7.71*	0.68	6.82	101	4.50	40.50	10.93	0.54	5.38
Light leisure activity	100	0.00	26.40	7.79*	0.45	4.54	101	3.30	19.80	6.27	0.31	3.12
Standing, carrying light loads	100	0.00	12.00	2.09	0.19	1.92	101	0.75	6.00	2.00	0.09	0.88
Reaching time from home to collage	100	0.56	13.50	4.33*	0.26	2.62	101	0.00	13.50	5.58	0.36	3.59
Low intensity aerobic exercise	100	0.00	8.48	1.27*	0.20	1.98	101	0.00	12.72	4.29	0.28	2.79
Planting, weeding etc.	100	0.00	5.18	0.55*	0.11	1.05	101	0.00	4.60	0.22	0.07	0.72
Colleeting water/wood	100	0.00	6.60	0.91*	0.15	1.49	101	0.00	13.20	1.00	0.22	2.18

(Contd...)

Non mechanised domestic chores	100	0.00	21.00	4.27*	0.33	3.31	101	0.00	5.60	1.23	0.14	1.39
Shopping	100	0.00	22.85	7.88*	0.59	5.90	101	0.00	13.71	3.78	0.28	2.81
Reading/studying	100	0.00	6.48	4.05	0.14	1.36	101	1.08	8.10	4.01	0.15	1.48
Watching TV	100	0.00	30.96	5.06*	0.39	3.92	101	0.00	10.32	4.07	0.20	2.00
Running small distance	100	0.00	12.70	3.14*	0.32	3.21	101	0.00	19.05	1.59	0.28	2.80
Swimming	100	0.00	18.00	0.88	0.29	2.89	101	0.00	18.00	0.89	0.27	2.74
Dancing	100	0.00	15.00	2.60*	0.36	3.57	101	0.00	10.00	0.54	0.15	1.52
Climbing stairs	100	0.00	24.44	5.30	0.48	4.83	101	0.00	18.33	4.90	0.29	2.96
Listening radio/music	100	0.00	4.29	2.25	0.14	1.37	101	0.00	8.58	2.39	0.14	1.36
Painting	26	0.32	2.00	1.26	0.21	0.67	15	0.64	3.18	1.04	0.27	0.82
Loading of goods on vehicle	100	0.00	13.05	0.40*	0.17	1.70	101	0.00	13.05	4.46	0.31	3.13
Ironing clothes	100	0.00	2.08	1.04*	0.06	0.62	101	0.00	1.66	0.51	0.04	0.44
Grooming animals	100	0.00	10.71	0.82	0.19	1.87	101	0.00	14.28	1.13	0.28	2.84
Running long distance	100	0.00	12.70	1.43	0.28	2.77	101	0.00	15.88	1.27	0.29	2.94
Sitting quietly	100	0.00	7.20	2.16	0.18	1.83	101	1.20	25.20	5.47	0.30	3.04
Gardening	100	0.00	5.32	0.98*	0.17	1.66	101	0.00	5.32	0.43	0.13	1.33
Washing hands/face	100	0.00	6.90	1.34*	0.09	0.90	101	0.00	2.30	0.98	0.05	0.47
Other activities	100	0.00	11.20	6.25	0.26	2.57	101	1.40	16.80	6.29	0.28	2.83
Talking with your family	100	0.35	7.00	2.92	0.14	1.38	101	0.70	7.00	3.01	0.15	1.51
Lying on bed	100	0.50	14.40	3.01	0.15	1.47	101	0.60	4.80	2.87	0.10	1.05

* Mean differ significantly between boys and girls.

is around 3 hours among the girls and 3.6 hours among the boys. Both the activities i.e. walking and time taken to reach school differ significantly between the sexes. The time spent for reading/studying is almost equal among the boys and girls and there is no statistical difference of this activity. It is 3.79 and 3.67 hours for girls and boys. The PAR of this activity also does not differ significantly among the sexes. Among the sports and recreational activities: lower intensity aerobic dance, watching TV, running small distances, swimming, listening radio/music, painting etc. are included. The time spent on these activities is 7.82 hours among the girls and 6.36 hours among the boys. The time spent on high physical activity level is significantly differs among the sexes such as low intensity aerobic dance, dancing. Besides, the time spent on watching TV is also differ significantly among the boys and girls.

The youth of Chamba also spent more time on other domestic activity and almost same time by the boys and girls (2.24 hours 2.23 hours respectively). The remarkable point is that both the boys and girls are involved in cooking and the time spent on it is 1.51 and 1.46 hours for boys and girls respectively. The youth also spent time on some general household works and obviously girls spent more time on it. Talking and gossiping is another activity performed by the youth and they spent more than two hours. Washing hands, face and personal care are also one of the important activities and they spent time on it which is more than an hour.

The PAL is one of the major determinants of health status. The calculated mean PAL value shows significantly differences among the boys and girls (Table 12.3). Interestingly, girls have significantly higher mean value than the boys. PAL range may also be classified in three different categories i.e. 1.40-1.69 which is sedentary or low activity level, 1.70-1.99 medium activity level and last is 2.0-2.40 which is active and vigorously active level. Hence the minimum value itself state that not a single individuals are under the category of light activity of life styles. The maximum value definitely shows vigorously active lifestyle but it is very difficult to maintain. The mean value of PAL comes under the category of vigorous or vigorously active lifestyle.

Table 12.3: Physical activity level of Chamba youth

Sex	*N*	*Range*	*Mean*	*SD*	*SE*
Girls	100	1.84- 2.88	2.21*	0.19	0.02
Boys	101	1.81-2.81	2.12	0.18	0.012

* Significantly differences between the means of boys and girls.

Disscusion

Adolescence is a period of rapid growth and behavioural changes, during which blood pressure, height, weight and body fat increase rapidly and physical activity level declines sharply. Recent changes in lifestyle have led to the global epidemic of high blood pressure levels. So, its regular monitoring and proper diagnosis of blood pressure levels are very important for developing healthy population of India.

Further studies of US Hispanics in comparison with other population sub-groups in 2013 also showed that weight change reflects the social environment of the community, pointing to the importance of social capital and interpersonal relationships in influencing weight status, especially among women. Brage *et al.*, (2004) found that students who engaged in a high amount of moderate physical activity under estimated weight. So, these studies reveal that increasing weight with age and height that up to certain age is due to the factors which depends upon the less physical activities, poor or diet with more fatty in nature. Physical fitness is often discussed with physical activity; it is a state or a condition which permits the individual to carry out her/his daily activities without undue fatigue and with sufficient reserve to enjoy active leisure pursuits. Hence, it can be said that the PAL of any population determined by their subsistence and other behavioural aspects. The Chamba youth spend highest time on resting activities. The PAL of the boys differs significantly from the girls and both the groups are under the category of high PAL.

Physical activity is a topic of current discussion, specifically in terms of health promotion and disease prevention. Physical inactivity, on the other hand, is indicated as a major risk factor for morbidity and mortality in adults (American College of Sports Medicine, 2001) and is increasingly being implicated as major risk factor for morbidity in children and adolescents (Strong *et al.*, 2005). Physical inactivity is also implicated in the recent worldwide epidemic of obesity (Caballero, 2007; WHO, 2000), which in turn is also an independent risk factor for morbidity and mortality in adults and also for metabolic morbidities in children and adolescents (Strong *et al.*, 2005).

Mortality rates from coronary heart disease increased dramatically after World War II, although the trend was already apparent early in the twentieth century (Smil, 1989; Stern, 1979). Mortality rates peaked in the 1960s in what was labelled the epidemic of coronary heart disease, and have declined since (Stern, 1979; Tyroler, 1988). The link between heart disease and physical inactivity was hypothesised in 1953 and physical inactivity as a risk factor for heart disease was established in 1992 (Paffenbarger *et al.*, 2001).

Besides, higher total and sport physical activity in men were associated with reduced risk for metabolic syndrome in young adulthood, while risks for individual metabolic syndrome components were reduced in women with increased total leisure, and sport physical activity. The study of Folsom *et al.*, (1996) stated the inverse relationship between physical activity and metabolic syndrome during young adulthood reported by other investigators. When assessing the activity levels of participants aged 20–38 years, investigators for the Bogalusa Heart Study found that low leisure-time activity was associated with greater values of factors implicated in the metabolic syndrome (insulin resistance index, waist circumference, total cholesterol to high density lipid ratio) after adjustment for demographic variables (Gustat *et al.*, 2002). Increased leisure time activity was associated with having fewer risk factors. Physical inactivity predicted increases in blood pressure and fasting insulin and glucose (Folsom *et al.*, 1996).

Physical activity level is determined by several factors: such as location i.e. urban or rural, terrain, availability of transport facility, economic status, home environment, cultural practices, availability of outdoor sport facility etc. The Chamba youth show very good status in comparison to other study conducted in India. Interestingly, girls have more PAL than the boys, which contradicting most of the Indian studies. The leisure activity pattern also favour higher PAL among the Chamba youth, whereas in other cases it is not favouring, especially, when time is spent for TV watching, lying on bed and sleeping (Kuriyan, 2007), gossiping etc. In conclusion, it can be said that the PAL level of the Chamba youth is very good and higher than the other studies. This is also supported by the prevalence of obesity level based on waist-stature ratio.

References

Anjana, RM., *et al.* 2014. Physical activity and inactivity patterns in India – results from the ICMR INDIAB study (Phase-1) [ICMR-INDIAB-5]. *International Journal of Behavioral Nutrition and Physical activity*. 11:26.

American College of Sports Medicine. 2001. Dose-response issues concerning physical activity and health: an evidence-based symposium. *Med Sci Sports Exerc*. 33(Suppl):S347-S641.

Aslesh, OP, Mayamol P., Suma RK., Usha K., Sheeba G., and Jayasree AK. 2015. Level of physical activity in population aged 16 to 65 years in rural Kerala, India. *Asia Pac J Public Health*. Aug 14. pii: 1010539515598835.

Brage, S., Wedderkopp N., Ekelund U., Franks PW., Wareham NJ., Andersen LB. and Froberg K. 2004. Features of the metabolic syndrome are associated with objectively measured physical activity and fitness in Danish children: the European Youth Heart Study (EYHS). *Diabetes Care*. 27: 2141-2148.

Caballero, B. 2007. The global epidemic of obesity: an overview. *Epidemiol Rev*. 29:1-5.

Folsom, AR., Jacobs DR. Jr, Wagenknecht LE., Winkhart SP., Yunis C., Hilner JE., Savage PJ., Smith DE., and Flack JM. 1996. Increase in fasting insulin and glucose over seven years with increasing weight and inactivity of young adults. The CARDIA Study. Coronary artery risk development in young adults. *Am J Epidemiol*. 144:235-246.

FAO/ WHO/UNU.2001. *Human energy requirements*: Report of a joint FAO/ WHO/ UNU Expert consultation. Rome.

Gustat, J, Srinivasan SR, Elkasabany A and Berenson GS.2002. Relation of self-rated measures of physical activity to multiple risk factors of insulin resistance syndrome in young adults: The Bogalusa Heart Study. *J. Clin. Epidemiol*. 55:997-1006.

Janson, I. and LeBlanc, AG. 2010. Review Systematic review of the health benefits of physical activity and fitness in school-aged children and youth. *International Journal of Behavioral Nutrition and Physical Activity*, 7:40.

Kuriyan, R., Bhat, S., Thomas, T., Vaz, M., and Kurpad, A.V. 2007. Television viewing and sleep are associated with overweight among urban and semi-urban South Indian children. *Nutr J*, 6, 25.

Paffenbarger, RS, Blair SN & Lee I-M. 2001. A history of physical activity, cardiovascular health and longevity: The scientific contributions of Jeremy N Morris, DSc, DPH, FRCP. *Int J Epidemiol*. 30:1184-1192.

Smil, V. 1989. Coronary heart disease, diet, and Western mortality. *Pop Dev Rev.* 15:399-424.

Stern, J M, Konner M, Herman T N & Reichlin S. 1986. Nursing behaviour, prolactin and postpartum amenorrhoea during prolonged lactation in American and Kung mothers. *Clinical Endocrinology*. 25: 247-258.

Strong, WB, Malina RM, Blimkie CJR, Daniels SR, Dishman RK, Gutin B, Hergenroeder AC, Must A, Nixon PA, Pivarnik JM, Rowland T, Trost S, Trudeau F. 2005. Evidence-based physical activity for school youth. *J Pediatr.* 146:732-737.

Tyroler, A. 1988. Introduction. In: Higgis MW, Luepker RW, editors. Trends in coronary heart disease: the influence of medical care. New York: Oxford. 3-15.

Vaz, M. and Bharathi, AV.2004. Perceptions of the intensity of specific physical activities in Bangalore, South India: implications for exercise prescription *JAPI*, 52.

WHO. 2000. Obesity: Preventing and Managing the Global Epidemic. Report of a WHO consultation.

WHO Technical Report Series No. 894. Geneva.

Theme – 5
Nutritional Status and Associated Factors

13

Nutritional Status and Associated Factors among the Tribal Children of Three Ecological Zones of India

—*Swapan Kumar Kolay and*
Gautam Kumar Kshatriya

Abstract

The objective of this study is to measure the nutritional status of preschool children, determine the factors that influence nutrition and estimate the degree of nutritional status. The data was collected from three ecological settings like Himachal Pradesh, Rajasthan and Gujarat respectively. For assessing the nutritional status of the children less than or equal to six years old belonging to Kinnaura, Bhil and Dhodia tribal population groups Himalayan, Desert and Coastal district in Himachal Pradesh, Rajasthan and Gujarat, a set of anthropometric measurements was undertaken on them. Anthropometric data were collected from the Anganwadi centres of different villages, the households and primary schools. The data were based on the cross-sectional sample of 1060 children ages less than one through six years. Out of these 1060 subjects, 519 were males and 491 females. In all, 12 body measurements were taken, viz. Crown heal length (upto two years)/ standing height, body weight, waist circumference, hip circumference, head circumference, chest circumference and mid upper arm circumference. It was found that household economic status, education of parents, prenatal care, visits of the mother (for access to health services), child's age, birth order and preceding birth interval are important determinants of child stunting. It was found that there exists a strong association between maternal and child nutritional status and maternal nutritional status and birth weight. This indicates that actions towards improving women and child nutrition should always be integrated for effective utilisation of scarce resources and to reduce the link (mother-child) of under0 nutrition. The children of very poor or low economic status households have the highest rates of malnutrition. This may be due to food insecurity in these households that negatively impacts the nutritional status of children in particular and the other household members in general. Therefore, measures should include government action to support the very poor, and to bring about rapid economic growth at the national level.

Key words: *Height, Weight, Stunting, Malnutrition, Pre-school, Himalays, Desert, Coastal.*

Introduction

Malnutrition is a serious problem, particularly for the poor and unprivileged. Children are most vulnerable to malnutrition due to low dietary intakes, inequitable distribution of food within the household, improper food storage and preparation, dietary taboos, infectious diseases and care. Prevalence of malnutrition has remained a problem of considerable magnitude in most of developing countries (Devi and Geervani, 1994). Malnutrition causes both emotional and physical suffering (Smith and Haddad, 2000) and is responsible for more than one half of all children's deaths worldwide (Pelletier *et al.*, 1995). Adults who sufer from malnutrition during their childhood are physically and intellectually less productive and suffer from higher levels of chronic illness and disability (Smith and Haddad, 2000).

Among rural pre-school children of the Kinnaura, Bhil and Dhodia prevalence of malnutrition among Himalayan, Desert and Coastal belt are remarkable. Empirical investigations have identified the problems of poverty and food insecurity which have prevailed among the low income population as well as high costs of living and dearth of animal protein among the causes of malnutrition (Kolay, 2012).

The objective of this study is to measure the nutritional status of pre-school children, determine the factors that influence nutrition and estimate the degree of nutritional status. It is expected that the findings from the study would serve as a guide to policy-makers, extension staff, food nutritionists and parents seeking to achieve some meaningful improvement in Kinnaura, Bhil and Dhodia children of three different ecological settings.

Material and Method

Data Collection

The data was collected from three ecological zones like, Himachal Pradesh, Rajasthan and Gujarat respectively. To achieve the goal only one district was chosen from each state. From each district three tehsils/ talukas were selected. Altogether nine tehsils/ taluka and 27 villages were selected to achieve the goal.

As per census (2001) the population of Himachal Pradesh was 60,77,248. Of this 15,02,170 persons are the Scheduled Tribes (STs) constituting 4.02 per cent of the total population of the state. Census 2011 had enumerated 84,298 Kinnaur of which male and female were 46,364 and 37,934 respectively. There was change of 7.61 per cent in the population compared to population of 2001. In the previous Census of India 2001, Kinnaur District recorded increase of 9.91 per cent to its population compared to 1991. The initial provisional data suggest a density of 13 in 2011 compared to 12 of 2001. Total area under Kinnaur district is of about 6,401 sq. km. With regards to sex ratio in Kinnaur, it stood at 818 per 1000 male compared to 2001 census figure of 857. The average national sex ratio in India is 940 as per latest reports of census 2011. The percentage of SC population to total population was 9.73 and that of ST 71.83. Together they constitute 82 per cent of the total population of the district. In the Himalayan region, the Kinnaur district had largest concentration of tribal population. Among the tribal population groups, Kinnaura constitute the largest population (55,973) of all tribal groups in Kinnaur (Census of India, 2001) and Bhot and Jad with each having strength of 130 and 74 respectively. Kinnaura were, therefore a logical choice as it was the largest Himalayan tribe of district Kinnaur.

On the other hand, as per Census (2001) the population of Rajasthan was 56,507,188. Of this 7,097,706 persons were the Scheduled Tribes

(STs) constituting 12.6 per cent of the total population of the state. In the western part of desert district of Rajasthan; the Bhil tribe of Barmer district was chosen for present study. Though, there are certain other districts of Rajasthan having a higher concentration of Bhils like Udaipur, Bhilwada, Baswara, Jalore etc, and the Bhils of Barmer were selected because of the desert condition.

The population of Gujarat in 2001 was 50,671,017. Of this 7,481,160 persons are the Scheduled Tribes (STs) constituting 14.8 per cent of the total population of the state (Census, 2001). In the coastal belt of Southern Gujarat, Surat and Valsad are two districts with largest concentration of tribal groups. Among the tribal population groups, Dhodia constitute the largest percentage (about 37 per cent) of all tribal groups in Valsad (3.59 lakhs, Census of India, 2001) and Bhil and Chandhuri together constitute 40.5 per cent of all the tribal population in Surat with each having a strength of 2.06 laks and 2.16 lacs respectively. Dhodia were, therefore a logical choice as it was the largest coastal tribe of district Valsad.

So the research work conducted on Kinnauras of Kinnaur, Bhils of Barmer and Dhodias of Valsad had brought important information on nutrition of pre-school children and their bio-cultural determinants. The quantum of qualitative and quantitative anthropometric and other data is presented in table 13.1.

Table 13.1: Quantum of qualitative, quantitative, anthropometric and other data collected during field work

District/ State/ Population (Sample Size)	*Qualitative (Sample Size)*	*Quantitative (No. of House-hold Schedule)*	*Anthropometric (No. of subjects)*
Kinnaur, Himachal Pradesh (Kinnaura)	FGD= 16, Case studies= 28	450	326
Barmer, Rajasthan (Bhil)	FGD= 9, Case studies= 27	450	352
Valsad, Gujarat (Dhodia)	FGD= 14, Case studies= 25	450	382
Total	**FGD= 39** **Case study=80**	**1350**	**1060**

Study Area

Kinnaur district is bounded on north by Spiti on the east by the Tibetan territory; on the south by Uttar Kashi district of Uttar Pradesh and on the west by Simla district (earlier Mahasu district). It is lies on both sides of Sutlej. It is situated between 31°05′ - 55′ and 32°05′ - 20′ north latitude and between 77°45′ - 0′ and 79°- 0′ - 50′ east longitude. It is stretched over 80 kms. in length and about 64 kms. in breadth. The total area of the district is 6401 sq. kms., which comprise of 11.5 per cent of the total area of the state of Himachal Pradesh. The district head-quarters is located at Recong Peo. There are three administrative sub-division in the district namely, Nichar, Kalpa and Pooh. Kalpa sub-division has two tehsils i.e. Kalpa and Sangla, Nichar sub-division has only one tehsil namely, Nichar and Pooh sub-division has two tehsils namely, Pooh and Morang. Information on 450 households has been collected from nine villages from three tehsils namely, Kalpa, Morang and Pooh.

Similarly, Barmer is one of the western part of desert district of Rajasthan. Barmer district is bounded on north by Jaisalmir district and on the south by Jhalor district and on the east by Palli and Jodhpur districts and on the west by Pakistan (international border), which is 270 kms. away from the district headquarters. It is situated between 24°58' and 26°32' north latitude and between 70°5' and 72°52′ east longitude. The total area of the district is 28387 sq. kms., which is the second largest district of Rajasthan. The data has been collected on 450 households from nine villages from three tehsils namely, Barmer, Chohtan and Ramsar.

Valsad district is bounded on north by Navsari district of Gujarat, on the east by the Maharashtra state, on the south by Dadra & Nagar Haveli and Maharashtra state and on the west by the Arabian Sea. It is situated between 20°07′ and 21°05′ north latitude and between 72°43′ and 73°00′ east longitude. The total area of the district is 5,244 Sq. kms., which comprise of 2.68 per cent of the total area of the state of Gujarat. Information on 450 households has been collected from nine villages from three talukas namely, Valsad, Pardi and Umargaon.

Assessment of nutritional status through anthropometric measurements

For assessing the nutritional status of the children less than one or equal to six years old belonging to Kinnaura, Bhil and Dhodia tribal population groups of Himalayan, Desert and Coastal district in Himachal Pradesh, Rajasthan and Gujarat, a set of anthropometric measurements was undertaken on them. Anthropometric data were collected from the Anganwadi centres of different villages, the households and primary schools. The data were based on the cross-sectional sample of 1060 children ages less than one through six years. Out of these 1060 subjects, 519 were males and 491 females. Salient findings of the present study are reported in the text below. In all, 12 body measurements were taken, viz. crown heal length (upto two years)/ standing height, body weight, waist circumference, hip circumference, head circumference, chest circumference and mid upper arm circumference.

The results of each of the 12 measurements mentioned above for both boys and girls are calculated. The statistical values are provided for each variable include mean, standard deviation and standard error of mean and the range. It is clearly evident from the analysis that irrespective of sex, there is even growth in each of the body measurements at successive ages. The t- values calculated for estimating the significance of the bisexual differences at different ages among Kinnaura, Bhil and Dhodia are presented in Table 13.2.

It can be seen from the table that in case of Kinnaura the bisexual differences, in general, are statistically non-significance for most of the measurement below three years. The differences become significant only after three years of age for various measurements. However, body weight, standing height and mid-upper arm circumference are the measurements, which do not show bisexual differences by and large.

It can thus be inferred that Kinnaura boys and girls under six years of age, in general, reveal similar pattern of growth for various body measurements in the present study. Besides numerous other contributory factors, cultural practices among Kinnauris are also responsible for these similarities. A girl child is never looked down upon in the society.

Table 13.2: Test of Significance (t-test) for Estimating Bi-Sexual Differences in Various Body Measurements at Different Ages Among the Kinnaura, Bhil and Dhodias

S. No.	Measurements	Age Group																	
		0-1			1-2			2-3			3-4			4-5			5-6		
		Kinnaura	Bhil	Dhodia	Kinnaura	Bhil	Dhodia	Kinnaura	Bhil	Dhodia	Kinnaura	Bhil	Dhodia	Kinnaura	Bhil	Dhodia	Kinnaura	Bhil	Dhodia
1.	Stature	1.35	5.28*	0.42	0.79	4.18*	0.35	2.66*	0.88	1.15	1.02	4.89*	0.29	1.93	6.04*	0.14	0.48	2.22*	0.03
2.	Body Weight	1.81	6.77*	0.47	0.80	6.62	2.66	0.70	2.39*	1.61	1.91	0.75	1.01	1.91	6.35*	0.39	0.34	2.0*	2.02
3.	Biceps Skinfold	0.64	1.82	0.58	1.41	4.80*	0.77	0.05	3.25*	0.69	1.0	1.88	0.85	1.37	7.25*	0.52	2.3*	3.73*	2.56
4.	Triceps Skinfold	1.38	0.41	0.18	0.03	6.09*	2.20	0.79	1.44	1.51	2.09	7.03*	2.05	1.73	2.47*	1.18	2.7*	3.76*	4.57
5.	Sub-Scapular Skinfold	1.02	15.0*	0	0.11	5.67*	1.86	0.23	7*	1.56	3.07	1	2.08	2.4*	0.05	0.82	2.1*	1.29	5.54
6.	Supra-Iliac Skinfold	1.55	2.66*	0.64	1.91	1.40	1.28	0.95	0.76	0.22	4.02*	3.89*	1.93	2.5*	8.07*	2.20	3.7*	2.1*	4.79
7.	Calf Skinfold	0	3.33*	1.2	0.91	1.27	0.93	0.56	0.44	0.07	1.78	10.4*	1.46	1.69	1.54	1.94	2.9*	2.36*	5.13
8.	Waist Circumference	1.71	4.33*	0.1	1.04	5.21*	0.59	0.94	5.02*	0.12	2.49	3.3*	0.4	1.65	0.53	0.41	0.49	1.47	0.17
9.	Hip Circumference	1.84	2.55*	0.13	1.58	2.45*	0.15	0.05	1.87	0.13	0.65	2.0	1.07	1.33	2.31*	1.27	1.33	2.79*	1.05
10.	Head Circumference	2.04*	2.30*	0.47	0.02	3.87*	3.57	2.54*	2.38*	3.81	3.14*	1.06	2.42	1.25	3.90*	0.52	2.0*	0.68	1.3
11.	Chest Circumference	2.24*	3.40*	0.19	1.45	5.0*	2.05	0	0.51	2.17	3.55*	1.74	1.92	3.35*	1.76	1.95	1.17	1.27	1.63
12.	Mid-Upper Arm Circumference	1.51	0.0	0.93	1.44	1.15	0.6	1.02	5.38*	0.13	0.27	9.06*	0.02	0.26	2.26*	0.64	0.44	1.40	1.12

*T- values are significant at 5 per cent level.

Since Kinnauries practice fraternal polyandry, therefore in the matter of sexuality women have enough freedom. There is no oppression and incidence of domestic violence is almost non-exist. As far as diet is concerned, both sons and daughters get equal preference. Kinnauri women enjoy respectable status in the society. The similar pattern of trend it also noticed in case of Dhodias.

On the other in case of Bhil, the bisexual differences, in general, are statistically significance for most of the measurement from one to six years of age. It can, thus be inferred that Bhil boys and girls under six years of age, in general, do not reveal similar pattern of growth for various body measurements in the present study. Besides numerous other contributory factors, cultural practices among Bhils are also responsible for these dissimilarities. Though there is no oppression and incidence of domestic violence, a girl child never looked like boy child in the society. As far as diet is concerned, both sons and daughters do not get equal preference, in general boys have to preference to meal and other things. Girls have also downloaded some restrictions.

Data on Indian population is available for six measurements on boys and girls of various ages an ICMR Technical Report Series No. 18 (1994) of which only four measurements, viz. body weight, standing height/ crown heel length, chest circumference and head circumference are comparable to the present study. Similarly, Ghosh (1992) provides data on additional measurements i.e. mid upper arm circumference for the children belonging to better socio-economic class.

Thus, a comparison of among Kinnaura, Bhil and Dhodia boys and girls has been made with the data available on boys and girls of rural India and of better socio-economic class. It can be seen that the pattern of growth of Kinnaura, Bhil and Dhodia children (boys and girls) in the five measurements taken for comparison with rural Indian children and children of better socio-economic class.

Protein calorie malnutrition (PCM) has been identified as a major health and nutritional problem in India. It appears among children in the early childhood, generally below six years. It is an important cause of morbidity and mortality in children and leads to impairment of physical and mental growth of the individuals who survive. The incidence of PCM in India in preschool age children is one to two per cent. A great majority of cases are in the category of mild to moderate in case of Kinnaura and Dhodia, but in Bhil it is moderate to severe form which frequently goes unnoticed. To identify children requiring nutritional and health intervention, a number of classifications have been proposed. Gomez classification which is weight for age percentage is based on weight retardation. Water lows classification defines two groups for PCM. Malnutrition with retarded growth, which is height for age percentage indicating stunting and malnutrition with low weight, which is weight for height indicating wasting.

Another indication of PCM is mid upper arm circumference, which is a reliable estimate of body's muscle mass. Between age one and five years, mid upper arm circumference hardly varies. Any arm circumference exceeding 13.5 cm. is sign of a satisfactory nutritional status, between 12.5 and 13.5 cm. it indicates mild to moderate malnutrition and below 12.5 severe malnutrition.

Table 13.3, 13.4 and 13.5 shows frequency of under nutrition according to various classifications. It can be seen from the tables that in case of Kinnaura about 45 per cent boys and 39 per cent girls suffer from mild to moderate forms of malnutrition according to weight for height where as 43 per cent of boys and 47 per cent of girls suffer from mild to moderate forms of undernutrition with respect to height for age. Similarly, 62 per cent of girls and 68 per cent of boys suffer from mild to moderate forms of malnutrition with respect to weight for age. Thus, weight loss is less pronounced as compare to stunting in girls.

On the other hand in case of Bhil about 15 per cent boys and 20 per cent girls suffer from moderate to severe forms of malnutrition according to weight for height whereas 18 per cent of boys and girls suffer from moderate to severe forms of undernutrition with respect to height for age. Similarly, 37 per cent of boys and 40 per cent of girls suffer from moderate to severe forms of malnutrition with respect to weight for age. In the present study girls are found to be more vulnerable as compare to boys in weight for age.

Similarly, in case of Dhodia about 59 per cent boys and 60 per cent girls suffer from mild to moderate forms of malnutrition according to weight for height whereas 58 per cent of boys and girls suffer from mild to moderate forms of under nutrition with respect to

Table 13.3: Assessment of Degree of Malnutrition according to Weight for Height

Nutritional Status	Boys			Girls			Total		
	Kinnaura (n= 177)	*Bhil (n= 167)*	*Dhodia (n= 174)*	*Kinnaura (n= 149)*	*Bhil (n= 176)*	*Dhodia (n= 158)*	*Kinnaura (n= 326)*	*Bhil (n= 343)*	*Dhodia (n= 332)*
Normal	54.24	38.32	37.20	60.40	36.39	35.92	57.06	35.86	54.24
Mild	40.68	46.70	43.29	32.22	43.18	36.62	36.81	44.89	40.68
Moderate	4.52	12.57	15.85	6.71	9.65	23.24	5.52	11.07	4.52
Severe	0.56	2.39	3.66	0.67	10.22	4.23	0.61	6.41	0.56

Table 13.4: Assessment of Degree of Malnutrition according to Height for Age

Nutritional Status	Boys			Girls			Total		
	Kinnaura (n= 177)	*Bhil (n= 168)*	*Dhodia (n= 174)*	*Kinnaura (n= 149)*	*Bhil (n= 184)*	*Dhodia (n= 158)*	*Kinnaura (n= 326)*	*Bhil (n= 352)*	*Dhodia (n= 332)*
Normal	55.37	29.16	39.63	47.65	32.60	42.25	51.84	30.96	40.85
Mild	35.59	52.97	43.29	38.93	49.45	44.37	37.12	51.13	43.79
Moderate	7.91	12.5	15.24	8.72	16.30	13.38	8.28	14.48	14.38
Severe	1.13	5.35	1.83	4.70	1.63	0.00	2.76	3.40	0.98

Table 13.5: Assessment of Degree of Malnutrition according to Weight for Age

Nutritional Status	Boys			Girls			Total		
	Kinnaura (n= 177)	*Bhil (n= 170)*	*Dhodia (n= 174)*	*Kinnaura (n= 149)*	*Bhil (n= 179)*	*Dhodia (n= 158)*	*Kinnaura (n= 326)*	*Bhil (n= 349)*	*Dhodia (n= 332)*
Normal	30.51	11.11	11.59	33.56	11.73	16.20	31.90	11.46	13.73
Mild	55.93	51.76	46.34	47.65	45.81	42.25	52.15	48.71	44.44
Moderate	11.86	29.41	37.20	14.09	33.51	35.21	12.88	31.51	36.27
Severe	1.70	7.64	4.88	4.70	8.93	6.34	3.07	8.30	5.56

height for age. Similarly, 77 per cent of girls and 84 per cent of boys suffer from mild to moderate forms of malnutrition with respect to weight for age. In the present study, boys are found to be more vulnerable as compare to girls in weight for age.

Moreover, mean mid upper arm circumferences in the present study among boys and girls are invariable above 13.5 cms. in age groups four and above for girls and two and above for boys indicating their satisfactory nutritional status.The entire ratios and mean upper arm circumference in the present study do not indicate any extraordinary situation regarding malnutrition.

We can estimate the prevalence of undernutrition in terms of weight for age and height for age and weight for height following the methodology proposed by the World Health Organisation. For each child Z scores for weight for age, height for age and weight for height were calculated. For estimating Z scores, the Anthro software package developed by the World Health Organisation was used and the WHO standard was adopted (WHO, 2010). A child having a Z score less than 2 was classified as undernourished in terms of either weight for age or height for age or weight for height. Finally, the prevalence of undernutrition was calculated as the proportion of children under three years of age who were classified as undernourished in terms of either weight for age or height for age or weight for height. Subsequently, the proportion of children who were stunted and wasted (SW), only stunted (S), only wasted (W) and the proportion of children who were neither stunted nor wasted was also calculated. As per the recommendation of the World Health Organisation that the nutritional status of the child should be assessed after taking into consideration both the height for age and the weight for height and not on the basis of the weight for age or weight for height alone. Following the recommendations put forwarded by the World Health Organisation, a child can be classified in any of the following four categories:

The child has low height for age as well as low weight for height (SW)	The child is stunted as well as wasted. The child is undernourished in both the dimensions of child nutrition.
The child has only low height for age (S)	The child is stunted but not wasted. The child is undernourished in one dimension of child nutrition.

The child has only low weight for height only (W)	The child is wasted but not stunted. The child is undernourished in one dimension of child nutrition.
The child has normal height for age as well as weight for height (N)	The child is neither stunted nor wasted. The child is not undernourished in any dimensions of child nutrition.

These children are apparently small but not unhealthy. But Bhil children belong to desert populations of Rajasthan. Further few severe cases of clinical signs of malnutrition like marasmus and protein calorie malnutrition have been recorded in the annual report of District hospital, Barmer district, Rajasthan. One must also consider that this group of tribal children belongs to Himalayan, Desert and Coastal populations, which has a special biological adaptation to its environment. Further no severe cases of clinical signs of malnutrition like marasmus and kwaskorhiorkor were observed in the present study. The prevalence of high frequency of mild to moderate among Kinnaura and Dhodia and in case of Bhil from moderate to severe forms of malnutrition can be attributed to infection like diarrhea, respiratory infections besides poor maternal health during and at the time of delivery and utilisation of health services among many. Cultural practices relating to environmental sanitation, personal hygiene, food habits and their attitudes towards modern medical care go a long way in elevating their sufferings.

Box: 1

Major Observation

- *Bisexual differences for 12 body measurements were not statistically significant for most of the measurements, revealing significant for most of the measurements, revealing similar pattern of growth among boys and girls of Dhodia and Kinnauras. However, bisexual differences among the Bhils are statistically significant for most of the measurements and growth pattern of boys are better than girls.*
- *A comparison of Dhodias and Kinnaura boys and girls with the comparable data on Indian rural boys from better socio-economic class did not reveal any great difference with them. But the Bhil boys and girls are lagging behind.*
- *At least 63 per cent of Bhils and 58 per cent of Dhodias boys and girls suffer from mild to moderate form of malnutrition and 89 per cent Kinnauras boys and girls fall in the normal to mild range for nutritional status according to weight for height and height for age.*
- *Mean and range of all the 12 measurements are supervisor among the Kinnauras is better than Dhodias and Bhils.*
- *Bhils are usually lagging behind in the overall nutritional status.*

Factors affecting Nutrition

Place of residence

Three districts of three different states were studied under present investigation namely, Kinnaur (Himachal Pradesh), Barmer (Rajasthan) and Valsad (Gujarat). Among these three districts, the distance of different village under study from district as well tehsil/ taluka headquarters is given in Table 13.6. The average distance of different villages from tehsil headquarters in Chohtan is 48 kms., followed by from Pooh tehsil it is 22 kms. and in Morang and Ramsar tehsil it is 19 kms. It can be observed from the Table 13.6 that the villages of Barmer and Kinnaur district are situated far off and in remote area. A comparative study showed that rural children are more likely to suffer from chronic energy deficiency than children in urban areas. These higher rates of rural malnutrition were also reported by local District Hospital, Kinnaur, Barmer and Valsad districts.

Household economic status

The pattern of annual household income obtained from different households on the present study is presented in Table 13.7. It can be clearly seen that more than 90 per cent of the Bhil households under study are categorised as poor followed by 50 per cent in Dhodia, whereas 50 per cent of the Kinnaura respondents had their annual income around one lakh rupees primarily because of the cultivation of apple orchards. As in the case of women, the economic status of a household is also one of the most important determinants of child nutritional status (UNICEF, 1990). Comparative studies on child nutrition for more than 15 countries (Sommerfelt *et al.*, 1994) and Bhil studies in showed that the higher the level of economic status of the household, the lower the level of child stunting.

Literacy rate, as evident from Table 13.8, female literacy in Dhodia was found to be 73.36 per cent which was 70 per cent in Kinnauras, whereas female literacy was found to be only 11.37 per cent in Bhils. Because of economic compulsion, drop out rate is very high after primary education in all the three tribal groups. Bhils are found lagging behind amongst all the three groups. Education is one of the most important resources that enable women to provide appropriate care for their children, which is an important determinant of children's growth and development (Engle and Menon, 1996). As per this study incidence of malnutrition decreased among young children with an increase in the level of mothers' education.

Table 13.6: Distance of villages from tehsil/ taluka headquarters

District	*Name of tehsil/ taluka*	*Distance of villages from tehsil/ taluka head-quarters (in kms)*		
		Average	*Maximum*	*Minimum*
Kinnaur	Pooh	22	30	0
	Morang	19	47	1
	Kalpa	7	13	1
Barmer	Barmer	16	25	7
	Chohtan	48	74	5
	Ramsar	19	30	5
Valsad	Valsad	9	18	4
	Pardi	13	35	3
	Umergaon	15	30	8

Table 13.7: Distribution of households as per annual income.

Category	*Percentage of Household*		
	Kinnaura (n= 450)	*Bhil (n= 450)*	*Dhodia (n= 450)*
<Rs 25,000 (Poor)	12.81	78.16	58.85
Rs 25,000- 60,000 (Lower middle)	27.44	14.21	28.77
Rs 60,000-1,50,000 (Moderate)	29.51	6.91	10.97
> Rs 1,50,000 (Prosperous)	30.24	0.72	1.41

Education of mother

The significance and direct relationship of the child's mother's education concurs with the assertion that maternal schooling is strongly associated with good child care and good health (Maxwell *et al.*, 2000). More education for women is associated with higher levels of household food availability, higher quality diets, better care practices and behaviours and better nutritional outcomes. This finding makes a good case for the use of educational empowerment and capacity building of women as a means of promoting food and nutritional status of children in particular and household members in general.

Table 13.8: Literacy Rate

	Percentage (< 6 yrs excluded)		
Literacy	***Kinnaura (n=1721)***	***Bhil (n= 1452)***	***Dhodia (n= 2012)***
Illiterate	20.06	73.36	17.45
Literate	79.94	26.64	82.55
Total	**100**	**100**	**100**
Literates			
Primary	33.51	67.62	29.43
Middle	26.85	21.21	27.92
High School	16.56	6.06	26.83
Higher Secondary	9.84	3.03	9.12
Graduate and Other	13.24	2.08	6.70
Total	**100**	**100**	**100**

Employment status of mothers

It can be seen from Table 13.9 that Kinnauras are distinguished by having apple orchards, 26 per cent of Dhodias are agriculturists, 74 per cent of Bhils work as labourers. Among the three tribal groups there is a great degree of variability in the occupation they practice. These include agriculture, government and private services, defence jobs, casual labour, business etc. Although women's employment enhances the household's accessibility to income, it may also have negative effects on the nutritional status of children, as it reduces a mother's time for childcare. Some studies have revealed that mothers of the most malnourished children work outside their home (Popkin, 1980; Abbi *et al.*, 1991). Another study argued that there is no association between maternal employment and children's nutritional status (Leslie, 1988).

Table 13.9: Types of Occupation

Type	*Percentage*		
	Kinnaura (n= 450)	*Bhil (n= 450)*	*Dhodia (n= 450)*
Labour	3.06	74.11	24.23
Pvt. Job	1.56	0.43	25.51
Agriculture	66.55	20.90	26.67
Govt. Job	22.66	2.30	13.30
Business	6.17	2.26	10.29
Total	**100**	**100**	**100**

Source of water and availability of toilet facility

In the present investigation it was found from Table 13.10 that in Valsad, boring water is the primary source of drinking water. In Basrmer, boring/tap water and water from well was found to be the primary source of drinking water. Whereas among Kinnaura the drinking water is directly obtain from natural resources *("Chasma, Ngasti, Nala or Khad")*.

Table 13.10: Sources of Drinking Water

Source	*Percentage of families users*		
	Kinnaura (n= 450)	*Bhil (n= 450)*	*Dhodia (n= 450)*
River/ Chasma	22.92	0.41	1
Boring/Tap shared	51.09	79.40	82.9
Tap (owned)	25.99	0.10	2.8
Well	-	20.08	14.2
Total	**100**	**100**	**100**

In the present study, it can be seen (Table 13.11) that open drainage system is found in Kinnauras. In Valsad and Barmer the drainage system is negligible which is a perpetual source of poor environmental hygiene and is chiefly responsible for mosquito breeding and also a potential source of fungal and bacterial infection leading to communicable diseases. Unfavourable health environment caused by inadequate water and sanitation can increase the probability of infectious diseases and indirectly cause certain types of malnutrition (UNICEF, 1990; Engle, 1992). This study showed that unprotected water source and non-availability of latrine were associated with low child stature.

Table 13.11: Disposal of Waste Water

Disposal Category	*Percentage of Houses*		
	Kinnaura (n= 450)	*Bhil (n= 450)*	*Dhodia (n= 450)*
Closed drainage	0.24	0.00	0.3
Open drainage	44.64	1.02	0.8
No drainage	55.12	98.98	98.9
Total	**100**	**100**	**100**

Child morbidity

Diarrhea and other infectious diseases manifested in the form of fever affect both dietary intake and utilisation, which may have a negative effect on improved child nutritional status. Crude Death Rate (CDR) has been estimated to be 10.00 for Kinnauras, 9.22 for Bhil and 7.8 for Dhodias, and 8.7 for Indian National Population and 7.7, 7.9, 8.5 for the population of Himachal Pradesh, Rajasthan and Gujarat respectively. Infant Mortality Rate (IMR) is a very sensitive indicator of mortality indicating a prenatal and post natal care of mother and infants. IMR were estimated as 74.1 for Kinnauras, 93.92 for Bhils and 72.0 for Dhodias per 1000 live births. 70/1000 live births for Indian National Population and 79/1000 live births for Rajasthan. IMR for Bhils, which is vary as compare to both Indian National Population and Rajasthan. A comparative study on children's nutritional status (Sommerfelt *et al.*, 1994) indicated that stunting was highest among children with recent diarrhea.

Childcare practices

Table 13.12 presents childcare practices among Kinnaura, Bhil and Dhodias. It can be seen from the table that most of the Dhodia (85 per cent) and Kinnaura (77 per cent) infants get their first feed as mother's milk with colostrums. Only 35 per cent Bhil infants get their first feed as mother milk with colostrums while 53 per cent infants get their first feed as butter, curd or poppy. Forty eight per cent of Dhodia infants and 75.75 per cent of Kinnaura infants receive their supplementary diet within six months, whereas only 2.39 per cent of Bhils infants receive supplementary diet within six months. Duration of breast feeding for about two years is highest for Bhils (90.63 per cent) followed by Dhodias (74.5 per cent) and Kinnauras (64.73 per cent). In the present context two years of breast feeding is a long duration and can be reasoned out from the fact that most of the Bhil women are involved in household activities and in those income generation activities, which do not require them to leave their kids at home.

Table 13.12: Childcare Practices

Childcare practices	*Percentage*		
	Kinnaura	*Bhil*	*Dhodia*
I. First feed given to infant	**(n=111)**	**(n= 395)**	**(n=130)**
a. Mothers milk with colostrums	77.18	34.69	84.9
b. Mothers milk without colostrums	19.93	11.71	12.0
c. Jaggery water	0.41	0.45	1.9
d. Any other	2.48	53.15	1.2
II. Age at first supplementary diet	**(n=97)**	**(n= 121)**	**(n=114)**
a. 6 months	75.75	2.39	48.0
b. 6-12 months	24.24	97.61	30.2
c. After one year	-	-	21.8
III. Duration of breast feed	**(n=69)**	**(n= 314)**	**(n=308)**
a. 1 year	20.98	0.91	6.7
b. 2 years	64.33	90.63	74.5
c. 3 years	9.80	8.07	16.5
d. More than 3 years	4.89	0.39	2.1

Age of child

Children's nutritional status is also more sensitive to factors such as feeding/weaning practices, care and exposure to infection at specific ages. A cumulative indicator of growth retardation (height-for-age) in children is positively associated with age (Anderson, 1995 as cited in Aschalew, 2000). Bhil children have also shown an increase in malnutrition with increase in age of the child.

Table 13.13: Status of Child Immunisation

Kinnaura		*Bhil*		*Dhodia*	
Vaccination	*per cent*	*Vaccination*	*per cent*	*Vaccination*	*per cent*
DPT (n= 108)		**DPT (n= 402)**		**DPT (n= 130)**	
Yes	97.45	Yes	30.14	Yes	95.00
No	2.55	No	69.86	No	5.00
Polio (n=106)		**Polio (n= 415)**		**Polio (n=128)**	
Yes	98.29	Yes	73.17	Yes	96.10
No	1.71	No	26.83	No	3.90
BCG(n= 108)		**BCG(n= 397)**		**BCG(n=130)**	
Yes	94.92	Yes	29.85	Yes	96.10
No	5.08	No	70.15	No	3.90
Measles (n= 99)		**Measles (n= 381)**		**Measles (n= 128)**	
Yes	75.45	Yes	28.84	Yes	69.90
No	24.55	No	71.16	No	30.10
Vit. A (n= 101)		**Vit. A (n= 316)**		**Vit. A (n= 127)**	
Yes	82.82	Yes	28.83	Yes	70.10
No	17.18	No	71.17	No	29.90

Immunisation status

Table 13.13 presents status of child immunisation among Kinnaura, Bhil and Dhodias. In the present study immunisation coverage was more than 90 per cent for DPT, Polio and BCG in Dhodia and Kinnauras but very poor was found to be among Bhils (less than 30 per cent for DPT and BCG and 73.17 per cent for Polio). The performance is not upto the mark and lot more is desired. Performance in administration of vaccination of DPT, BCG, measles and vitamin-A prophylaxis needs further strengthening of efforts in this direction. Incidentally government institution is the primary source of vaccination in more than 70 per cent of cases. By and large coverage is not satisfactory especially in Bhils of Barmer district. Lack of literacy, awareness and slackness among these tribal groups were the main reasons for not immunising their children.

Birth order

It has been found that Bhil parents give less attention to older children in comparison to Kinnauras and Dhodias, when they give birth to a new child who needs much attention and care. Crude birth rate has been estimated to be 21.6 for Kinnauras, 34.05 for Bhils and

24.5 for Dhodias. Whereas 26.1 for India and 22.6, 31.4 and 25.4 for Himachal Pradesh, Rajasthan and Gujarat respectively. The estimated CBR is higher than the CBR of the state of Rajasthan and Indian National Population. Total fertility rate of Kinnaura, Bhil and Dhodia women in various child bearing age groups indicate a total fertility of 2.91 for Kinnauras, 5.77 for Bhils and 3.86 for Kinnauras estimated from the average parity of women in the age group 45 through 49 years. General fertility rate for Kinnauras, Bhils and Dhodias was estimated as 162, 192.9 and 123 respectively. For both TFR and GFR Bhils were showing higher values than Indian National Population. One study showed that stunting is rare in birth order 2-3 (Sommerfelt *et. al.*, 1994) and higher birth order (5+) is positively associated with child malnutrition (Jeyaseelan, 1997).

Birth interval of the child

Closely spaced pregnancies are often associated with the mother having little time to regain lost fat and nutrient stores (ACC/SCN, 1990). Higher birth spacing is also likely to improve child nutrition, since the mother gets enough time for proper childcare and feeding. Studies in developing countries showed that children born after a short birth interval (less than 24 months) have higher levels of stunting.

Interrelationship between maternal and child nutrition

Birth weight, child growth and adolescent growth determine nutritional status before and during pregnancy (maternal nutrition). Maternal nutrition also influences foetal growth and birth weight (ACC/SCN, 1992). The presence of an intergenerational link between maternal and child nutrition means a small mother will have small babies who in turn grow to become small mothers. Some findings on the relationship between maternal and child nutrition (Loaiza, 1997; Teller *et al.*, 2000; Genebo *et al.*, 1999) showed that a high proportion of low birth weight and stunted children were observed among malnourished mothers.

Conclusion and Policy Implications

It was found that household economic status, education of parents, prenatal care, visits of the mother (for access to health services), child's age, birth order and preceding birth interval are important determinants of child stunting.

This study arrives at the following conclusions to improve women and children nutritional status. Most of the socio-economic variables affecting the nutritional status of children. It was also found that there exists a strong association between maternal and child nutritional status and maternal nutritional status and birth weight. This indicates that actions towards improving women and child nutrition should always be integrated for effective utilisation of scarce resources and to reduce the link (mother-child) of undernutrition.

This study revealed that children of very poor or low economic status households have the highest rates of malnutrition. This may be due to food insecurity in these households that negatively impacts the nutritional status of children in particular and the other household members in general. Therefore, measures should include government action to support the very poor, and to bring about rapid economic growth at the national level. It is important to develop community based interventions giving priority to very poor households as a short term solution. Urgent implementation of poverty reduction strategies and programmes designed by the government could also serve as a long term solution to the problem.

It should be noted that over 88 per cent of Bhils and nearly 30 per cent of Kinnaura and Dhodia women reported having no education. It is therefore necessary to promote universal education of girls and women. The results showed that education of parents is one of the important determinants of children's nutritional status. Children of educated parents are at a lower risk of malnutrition, if the risks observed for other variables are eliminated. This indicates that parents who receive even a minimum basic education (even in the poor households) are generally more aware than those who are not educated of the need to utilise available resources for the improvement of the nutritional status of their children. It is therefore imperative that young girls and boys be enrolled in compulsory primary school education and opportunities should also be given to adult women and men to take part in non-formal education. Health and nutrition education should also be an integral part of the education process.

Close spacing of births, i.e. having a preceding birth interval of less than 24 months, showed a significant nutritional deficit in the younger children, particularly in the rural areas of Barmer district of Rajasthan. This may be associated with risk factors such as mothers' inadequate capacity for caring for her children. The mother herself may be biologically depleted from too frequent births and this could also negatively affect the nutritional status of the newborn baby. Therefore, access to services for child spacing could benefit the youngest child and the mother. Prolonging the intervals between births, through increasing demand for family planning and/or fulfilling unmet need for family planning, could be important elements of strategies to improve child nutrition.

This study has also indicated that exclusive breastfeeding upto 6 months of age is not widely practiced nor is the timely introduction of weaning foods at about six months. Therefore, education with this regard is also important intervention.

Therefore, further research on socio-cultural practices, intra-household food distribution, women's workload, seasonal food insecurity and other related factors is suggested.

References

Abbi, R., P. Christian, S. Gujral, and T. Gopaldas. 1991. The impact of maternal work on the nutrition and health status of children. *Food and Nutrition Bulletin* 13(1):20-24.

Administration Committee on Coordination–Sub-Committee on Nutrition (ACC/SCN). 1990. *Women and nutrition.* Symposium report, Nutrition Policy Discussion Paper No. 6.

Administration Committee on Coordination–Sub-Committee on Nutrition (ACC/SCN). 1992. *Second report on the world nutrition situation,* Vol. 1 and 2, *Global and regional results*, New York.

Aschalew, G. 2000. Determinants of nutritional status of children in Amhara Region: A case study of Misrak Gojjam and Semen Wello Zones. M.Sc Thesis, DTRC/IDR, AAU.

Census of India. 2001. Ministry of Home Affairs, New Delhi.

Devi, P.Y. and P. Geervani, 1994. Determinants of nutritional status of rural preschool children in Andhra Pradesh, India, Food Nutr. Bull., 15: 335-342. www.unu.edu/unupress/food/8F154e/8F154 E0c.htm.

Engle, P.L. 1992. Care and child nutrition. Theme paper for the international conference on nutrition (ICN): Paper prepared for nutrition section, UNICEF, New York.

Engle, P.L., and P. Menon. 1996. *Urbanisation and care giving: Evidence from south and eastern Africa*. San Luis, California: Department of Psychology and Human Development, California Polytechnic, Stat. University. pp. 4-24.

Genebo, T., W. Girma, J. Hadir, and T. Demmissie. 1999. The association of children's nutritional status to maternal education in Ziggbaboto, Guragie Zone South Ethiopia. *Ethiopian Journal of Health Development* 13(1):55-61.

Ghosh, S. 1992. The feeding and care of infants and young children, VHAI, New Delhi.

ICMR. 1984. ICMR Technical Report series no. 78, ICMR, New Delhi.

Jeyaseelan, L. 1997. Risk factors for malnutrition in south India children. *Journal of Biosocial Science* 1: 93-100.

Kolay, S.K. 2012. *"Determinants of Nutritional Status among the Dhodia in Coastal Region"*, In: S. N. Chaudhary (ed), Tribal Health and Nutrition, Rawat Publication, New Delhi.

Kshatriya, G.K. 2000. Ecology and Health with special reference to Indian tribes, In Man-Environment relationship eds. M.K. Bhasin and V. Bhasin, Delhi, pp. 229-245.

Leslie, J. 1988. Women's work and child nutrition in the third world, *World Development* 16(1).

Loaiza, Edilberto. 1997. *Maternal nutritional status*. DHS Comparative Studies No. 24, Calverton, Maryland, USA: Macro International Inc.

Maxwell, D., C. Levin, M. Annar-Klemesu, M. Ruel, S. Morris and C. Ahiadeke. 2000. Urban Livelihoods and Food and Nutrition Security in Greater Accra, Ghana. Research Report 112. International Food Policy Research Institute (IFPRI) Washington, D.C., USA.

Morbidity Report. 2003-4. Barmer District Hospital, District Barmer, Rajasthan.

Pelletier, D.L., E.A. Frongillo, D.D. Schroeder and J.P. Habicht. 1995. The effects of malnutrition on child mortality in developing Countries. *Bull. W. Health Org.*, 73: 443-448.

Popkin, B.M. 1980. Time allocation of the mother and child nutrition. *Ecology of Food and Nutrition* 9:1-14.

Sadhukhan, S.K, C. Chatterjee, P. Shrivastava, J.C. Sardar, G.K. Joardar, S. Lahiri. 2010. Validity of mid arm circumference to detect protein energy malnutrition among 8-11 months old infants in a rural medical college of West Bengal, *Journal of Indian Medical Association*, Kolkata, Sept, 108(9):559-62.

Smith, L.C. and L. Haddad, 2000. Explaining Child Malnutrition in Developing Countries: A Cross Country Analysis. IFPRI. Washington, DC: International Food Policy Research Institute. Research Report 111.

Sommerfelt, A. Elizabeth, and S. Kathryn. 1994. *Children's nutritional status.* DHS Comparative Studies No. 12. Calverton, Maryland, USA: Macro International Inc.

Teller, H., and G. Yimar. 2000. Levels and determinants of malnutrition in adolescent and adult women in southern Ethiopia. *Ethiopian Journal of Health Development* 14(1):57-66.

UNICEF. 1990. *Strategies of improving nutrition of children and women in developing countries*, New York.

Verma, I.C., Kaur, M. and Das, G.P. 1994. Laboratory Manual for health professionals in tribal/ rural areas, Genetic unit, AIIMS, New Delhi.

World Health Organisation. 2010. *WHO Anthro for Personal Computers. Software for Assessing the Growth and Development of the Worlds Children, Version 3.1, 2010,* Geneva, World Health Organisation.

World Health Organisation. 2006. *WHO Child Growth Standards, Length/ Height for Age, Weight for Age, Weight for Length, Weight for Height and Body Mass Index for Age, Methods for Development,* Geneva, World Health Organisation, Department of Nutrition for Health and Development.

14

Nutritional Status of Santal Children in Two Villages of Birbhum District, West Bengal

—*Jyoti Ratan Ghosh*

Abstract

The aim of the present study was to report the prevalence of undernutrition in terms of stunting, wasting and underweight among Santal tribal children in Birbhum district of West Bengal, India. A total of 272 (boys=138, girls=134) Santal children aged between 2-18 years were studied. Height and weight measurements were taken following standard techniques. Three commonly used undernutrition indicators, that is, stunting, underweight and wasting were used to evaluate the nutritional status of the subjects, by using the internationally accepted National Centre for Health Statistics (NCHS) standard as reference population. The overall prevalence of stunting, wasting and underweight were 46.4 per cent, 51.4 per cent and 32.6 per cent, respectively in boys and 49.3 per cent, 40.3 per cent and 35.6 per cent, respectively in girls. However, there were no significant sex differences in the three nutritional indicators. The prevalence of stunting and underweight was higher in monthly family income group d" Rs 2500.00, compared with monthly family income group e" Rs 2501.00. In conclusion, the Santal children revealed poor nutritional status with a higher incidence of wasting in boys and stunting in girls. Further studies are needed among a larger sample for effective planning of nutritional intervention programmes.

Introduction

Tribals are one of the most exploited and deprived sections of the population in the Indian society. In all indicators of development, they remain the most excluded despite the fact that various kinds of policies and programmes have been pursued for their upliftment in the post-independence India (Xaxa, 2011). Scheduled tribe people constitute about 8.6 per cent of the total population in India and 5.8 per cent of the total population of West Bengal (Census, 2011). They live in unique physical, socio-economic and cultural environment, isolated from general population. In view of their habitat and food habits, they form a distinct group compared to other populations (Rao *et al.*, 2006).

The level of child undernutrition remains unacceptable throughout the world, with 90 per cent of the developing world's chronically undernourished (stunted) children living in Asia and Africa (UNICEF 2013). More than one third of the world's children who are wasted live in India. India accounts for more than three out of every ten stunted children in the world. Undernutrition is substantially higher in rural than in urban areas and children from scheduled tribes have the poorest nutritional status (UNICEF, 2013). Undernutrition is a source of major waste of resources and lost productivity, because children who are undernourished are less productive both physically and intellectually as adults (Chakrabarty and Bharati, 2010).

Malnutrition in terms of undernutrition or overnutrition is the condition arising due to the intake of inappropriate nutrients in diet and choosing an appropriate nutritional status measure is a complex decision based on objective data collection (Chakrabarty and Bharati, 2010). To combat this problem, anthropometric measurements is the most commonly used because this method is simple and inexpensive compared to other existing methods (FAO, 1996; Chakrabarty and Bharati, 2010). Undernutrition in children is conventionally determined through measurement of height, weight and age. The most commonly used indices derived from these measurements are stunting (low height-for-age), wasting (low weight-for-height) and underweight (low weight-for-age) (WHO 1995; Nandy *et al.*, 2005). However, there is little information on the nutritional status of pre-adolescent and adolescent tribal children in India (Bisai *et al.*, 2008; Chakrabarty and Bharati, 2010). In view of the above, the aim of the present study was to report the prevalence of undernutrition in terms of stunting, wasting and underweight among Santal tribal children in Birbhum district of West Bengal, India.

Materials and Methods

The present community-based cross-sectional study was undertaken in two Santal villages of Birbhum district, about 150 km from the Kolkata city. A total of 272 (boys=138, girls=134) Santal children aged between 2-18 years were assessed. Information on ethnicity, age, gender and monthly family income was collected on a pre-tested questionnaire. Children's age was recorded as reported by mother and verified further with other senior members of the family. Height and weight measurements were taken on each subject following the standard techniques (Lohman *et al.*, 1988). Height was measured to the nearest 0.1 cm using moveable anthropometer. Weight was measured to the nearest 0.5 kg using weighing machine.

Three commonly used undernutrition indicators, that is, stunting, underweight and wasting were used to evaluate the nutritional status of the subjects, by using the internationally accepted National Centre for Health Statistics (NCHS) (Hamill *et al.*,1979; WHO, 1983) standard as reference population. Stunting was the proportion of children below -2 standard deviation (SD) from median height-for-age National Centre for Health Statistics/ World Health Organisation (NCHS/WHO) reference population. Underweight is the proportion of children below -2 SD from median weight-for-age of the NCHS/WHO reference population. Wasting is the proportion of children below -2 SD from the median weight for height of the NCHS/WHO reference population. Severe and moderate undernutrition was the proportion of children < -3 SD and between -3 SD - <-2 SD, respectively. However, for the assessment of underweight, we considered only boys (95 boys) below 145 cm and girls (90 girls) below 137 cm, because of the lack of NCHS/ WHO reference value for weight by stature above 145 cm in boys and 137 in girls.

Student's t-test were undertaken to test for sex differences in age, monthly family income, height and weight. Chi-square test was performed to test for differences in prevalence. We used 50th percentile as cut-off value to categorise monthly family income in two groups.

Statistical analyses were performed using the SPSS package. A p-value of <0.05 was considered as significance.

Results

A total of 272 children, 138 (50.74 per cent) boys and 134 (49.26 per cent) girls were studied. The mean age of the subjects were (boys: mean = 10.33 years, SD = 4.9 years; girls: mean = 10.07 years, SD = 5.06 years) similar. There was no significant age difference (p> 0.05) between them. The mean and standard deviation of age, height and weight are presented in Table 14.1. The mean height and weight of Santal boys were 128.45 (SD 24.82 cm) cm and 26.72 (SD 12.64 kg) kg, respectively. Among Santal girls these were 123.88 (SD 23.29 cm) cm and 27.06 (SD 14.16 kg) kg, respectively. The mean monthly family income of the Santal boys and girls were Rs 2799.93 (SD Rs 1988.80) and Rs 2907.84 (SD Rs 2362.11) respectively. There was no significant (p>0.05) sex differences in height and monthly family income. However, weight was significantly (p<0.05) higher in Santal girls compared to boys.

Table 1: Characteristics of the studied population

Variables	*Boys*		*Girls*	
	Mean	*SD*	*Mean*	*SD*
Age (year)	10.33	4.9	10.07	5.06
Height (cm)	128.45	24.82	123.88	23.29
Weight (kg)*	26.72	12.64	27.06	14.16

*significant difference; p<0.05

Nutritional status of the Santal boys and girls are presented in Table 14.2. The overall (severe + moderate) prevalence of stunting, wasting and underweight were 46.4 per cent, 51.4 per cent and 32.6 per cent, respectively in boys and 49.3 per cent, 40.3 per cent and 35.6 per cent, respectively in girls. It is important to note that the prevalence of severe stunting (51.5 per cent vs. 40.6 per cent) and underweight (53.1 per cent vs. 38.7 per cent) was higher in girls compared with boys. However, the prevalence of severe wasting was higher (35.2 per cent vs. 29.6 per cent) in boys compared with girls.

Contrary to that, the prevalence of moderate stunting (59.4 per cent vs. 48.5 per cent) and underweight (61.3 per cent vs. 46.9 per cent) was higher in boys compared with girls. The prevalence of moderate wasting was higher (70.4 per cent vs. 64.8 per cent) in girls compared with boys. However, there were no significant sex differences in the three nutritional indicators.

Table 14.2: Nutritional status of the studied population

Nutritional indicators		*Severe*	*Moderate*	*Undernutrition*	*Normal*
Boys					
	Stunting	26 (40.60)	38 (59.40)	64 (46.40)	74 (53.60)
	Wasting	25 (35.20)	46 (64.80)	71 (51.40)	67 (48.60)
	Underweight	12 (38.70)	19 (61.30)	31 (32.60)	64 (67.40)
Girls					
	Stunting	34 (51.50)	32 (48.50)	66 (49.30)	68 (50.70)
	Wasting	16 (29.60)	38 (70.40)	54 (40.30)	80 (59.70)
	Underweight	17 (53.10)	15 (46.90)	32 (35.60)	58 (64.40)

Table 14.3: Nutritional status by monthly family income and sex

Monthly family income	*Nutritional indicators*	*Severe*	*Moderate*	*Undernutrition*	*Normal*
Boys					
≤Rs 2500.00					
	Stunting	14 (53.85)	27 (71.05)	41 (64.06)	42 (56.76)
	Wasting*	9 (36.00)	36 (78.26)	45 (63.38)	38 (56.72)
	Underweight	8 (66.67)	10 (52.63)	18 (58.06)	41 (64.06)
≤Rs 2501.00					
	Stunting	12 (46.15)	11 (28.95)	23 (35.94)	32 (43.24)
	Wasting*	16 (64.00)	10 (21.74)	26 (36.62)	29 (43.28)
	Underweight	4 (33.33)	9 (47.37)	13 (41.94)	23 (35.94)
Girls					
≤Rs 2500.00					
	Stunting	23 (67.65)	25 (78.13)	48 (72.70)	39 (57.35)
	Wasting	11 (68.75)	24 (63.16)	35 (64.81)	52 (65.00)
	Underweight*	8 (47.06)	12 (80.00)	20 (62.50)	44 (75.86)
≤Rs 2501.00					
	Stunting	11 (32.35)	7 (21.87)	18 (27.30)	29 (42.65)
	Wasting	5 (31.25)	14 (36.84)	19 (35.19)	28 (35.00)
	Underweight*	9 (52.94)	3 (20.00)	12 (37.50)	14 (24.14)

Nutritional status by monthly family income and sex are presented in Table 14.3. The result revealed that in both sexes the overall prevalence of undernutrition in terms of stunting (boys, 64.06 per cent vs. 35.94 per cent; girls, 72.70 per cent vs. 27.30 per cent), wasting (boys, 63.38 per cent vs. 36.62 per cent; girls, 64.81 per cent vs. 35.19 per cent) and underweight (boys, 58.06 per cent vs. 41.94 per cent; girls, 62.50 per cent vs. 37.50 per cent) was higher in monthly family income group ≤ Rs 2500.00, compared with monthly family income group ≥ Rs 2501.00. Moreover, the prevalence moderate stunting (boys, 71.05 per cent vs. 28.95 per cent; girls, 78.13 per cent vs. 21.87 per cent), wasting (boys, 78.26 per cent vs. 21.74 per cent; girls, 63.16 per cent vs. 36.84 per cent) and underweight (boys, 52.63 per cent vs. 47.37 per cent; girls, 80.00 per cent vs. 20.00 per cent) was also higher in monthly family income group ≤ Rs 2500.00, compared with monthly family income group ≥ Rs 2501.00, in both boys and girls. However, the prevalence of severe stunting (53.85 per cent vs. 46.15 per cent) and underweight (66.67 per cent vs. 33.33 per cent) was higher in monthly family income group ≤ Rs 2500.00, compared with monthly family income group ≥ Rs 2501.00, in boys. On the other hand the prevalence of severe wasting (64.00 per cent vs. 36.00 per cent) was significantly ($p<0.05$) higher in monthly family income group ≥ Rs 2501.00, compared with monthly family income group ≤ Rs 2500.00. In girls, the prevalence of severe stunting (67.65 per cent vs. 32.35 per cent) and wasting (68.75 per cent vs. 31.25 per cent) was higher in monthly family income group ≤ Rs 2500.00, compared with monthly family income group ≥ Rs 2501.00. But, the prevalence of severe underweight (52.94 per cent vs. 47.06 per cent) was significantly ($p<0.05$) higher in monthly family income group ≥ Rs 2501.00, compared with monthly family income group ≤ Rs 2500.00.

Discussion

Undernutrition among children and adolescent are considered as a serious public health problem in developing countries such as in India, where the vast majority of the populations are undernourished and underprivileged (Singh and Mondal, 2013). It is well known that in India there is a wide ethnic heterogeneity. Thus, an assessment of nutritional status of each ethnic group is important for improvement of their nutrition and health status, as well as for the overall development of the community. This is because of the fact that, nutrition is a core pillar of human development. Large-scale programming not only can reduce the burden of undernutrition and deprivation in countries but also can advance the progress of nations (UNICEF, 2013). However, the result of the present study to understand the prevalence of undernutrition in terms of stunting, wasting and underweight among Santal children in Birbhum district of West Bengal, India, revealed that the overall prevalence of stunting, wasting and underweight in the study population was 47.79 per cent, 45.95 per cent and 34.05 per cent, respectively. The prevalence of severe under-nutrition i.e. stunting, wasting and underweight was 22.06 per cent, 15.07 per cent and 15.67 per cent respectively. Comparison of prevalence of undernutrition in the present study with the Santal children of Puruliya District revealed a higher prevalence of stunting, wasting and underweight in the present study (Chowdhury *et al.*, 2008). Comparison of prevalence of undernutrition with Lodha children also revealed a higher prevalence of stunting, wasting and underweight in the present study (Bisai *et al.*, 2008).

In order to understand the prevalence of undernutrition in boys and girls, the result revealed that the prevalence of stunting, wasting and underweight were 46.4 per cent, 51.4 per cent and 32.6 per cent, respectively in boys and 49.3 per cent, 40.3 per cent and 35.6 per cent, respectively in girls. Thus, the overall prevalence of undernutrition in terms of stunting and wasting was higher in girls. But, the prevalence of moderate stunting and moderate underweight was higher in boys. The cardinal features of the present study were the higher prevalence of severe stunting and underweight in girls and the prevalence of wasting and severe wasting in boys. Comparatively higher prevalence of severe stunting and underweight was also observed in Santal in girls of Puruliya District (Chowdhury *et al.*, 2008). A recent study in Bhil tribal children of Madhya Pradesh also revealed comparatively higher prevalence of stunting and wasting in girls (Pradhan and Sharma, 2011). Contrary to that, study in Kora-Mudi children of Paschim Medinipur revealed comparatively higher prevalence of stunting and wasting in boys (Bisai and Mallick, 2011). Higher prevalence of severe stunting and underweight in girls might be due to the chronic and acute undernutrition, which is the result of prolonged food deprivation and/ or disease or illness. On the other hand, higher prevalence of wasting in boys of the present study might be an indicator of acute undernutrition, the result of more recent food deprivation or illness (Nandi *et al.*, 2005). Study also showed that there was a close association between adolescent undernutrition, maternal malnutrition, and low birth weight (Rao *et al.*, 2006).

However, when we used 50th percentile as cut-off value to categorise monthly family income in two groups to understand its association with nutritional status, the result revealed that in both sexes, the overall and moderate prevalence of stunting, wasting and underweight were higher in lower income group. Moreover, the prevalence of severe stunting and severe underweight in boys and severe stunting and severe wasting in girls were also higher in lower income group. In a recent study Walingo and Kidake (2013) also observed that households socioeconomic status and income levels were the major determinants of the nutrition outcome of preschool children in Vihiga. Another study in Nepal also observed significant association of income with malnutrition in school children (Joshi *et al.*, 2011). A similar result was also observed by Babar *et al.*, (2010). The children from households with a low or very low socioeconomic status had 2.5 times the risk of being underweight relative to children who came from households with middle to upper socioeconomic status (Kikafunda *et al.*, 1998; Babar *et al.*, 2010). Because household's dietary diversity and intake of energy reduces when food prices increase and incomes decrease. The reduction of food and nutrient consumption increases the severity of nutrient deficiencies and increases the prevalence of malnutrition (Sari *et al.*, 2010; Walingo and Kidake, 2013). However, low levels of nutrition adversely affect physical and mental growth of children. Malnutrition in early childhood is associated with significant functional impairment in adult life, reduced work capacity and decreasing economic productivity (Delpeuch *et al.*, 2000; Babar *et al.*, 2010).

Thus, the present study in Santal children revealed poor nutritional status with a higher incidence of wasting in boys and stunting in girls. However, one of the limitations of the present study is the small sample size. Further studies are needed among a larger sample for effective planning of nutritional intervention programmes.

Acknowledgements

All participating subjects and villagers are gratefully acknowledged for their cooperation. Authors also acknowledge Visva-Bharati, Govt. of India for providing partial financial assistance.

References

Babar, NF, Muzaffar R, Khan MA, Imdad S. 2010. Impact of socioeconomic factors on nutritional status in primary school children. *J Ayub Med Coll Abbottabad*. 22: 15-18.

Bisai, S, Bose K, Ghosh A. 2008. Prevalence of undernutrition of Lodha children aged 1-14 years of Paschim Medinipur district, West Bengal, India. Iran *J Pediatr*. 18: 323-329.

Bisai, S, Mallick C. 2011. Prevalence of undernutrition among Kora-Mudi children aged 2-13 years in Paschim Medinipur District, West Bengal, India. *World J Pediatr*. 7:31-36.

Chakrabarty, S, Bharati P. 2010. Nutritional status among the Shabar tribal children living in urban, rural and forest habitats of Orissa, India. *IJPH*. 7: 3-10.

Chowdhury, SD, Chakraborty T, Ghosh T. 2008. Prevalence of undernutrition in Santal children of Puruliya district, West Bengal. *Indian Pediatr*. 45:43-46.

Delpeuch, F, Traissac P, Martin-Pre Y, Massamba JP, Maire B. 2000. Economic crisis and malnutrition: socioeconomic determinants of anthropometric status of preschool children and their mothers in an African urban area. *Public Health Nutr*. 3:39-47.

Food and Agriculture Organisation (FAO). Sixth world food survey. Rome: Food and Agriculture Organisation, 1996.

Hamill, PV, Drizd TA Johnson CL, Reed RB, Roche AF, Moore WM. 1979. Physical growth: national center for health statistics per centiles. *Am J Clin Nutr*. 32: 607-629.

http://www.tribal.gov.in/WriteReadData/CMS/Documents/201306110208002203443DemographicStatusofScheduledTribePopulationofIndia.pdf, accessed on 15.07.2013

Joshi, HS, Gupta R, Joshi MC, Mahajan V. 2011. Determinants of nutritional status of school children : a cross-sectional study in the Western region of Nepal. *NJIRM*. 2: 10-15.

Kikafunda, JK, Walker AF, Collett D, Tumwine JK. 1998. Risk factors for early childhood malnutrition in Uganda. *Pediatrics*. 102:e45.

Lohman, TG, Roche AF, Martorell R. 1988. Anthropometric standardisation reference manual. Human Kinetics Books, Chicago.

Nandy, S, Irving M, Gordon D, Subramanian SV, Smith GD. 2005. Poverty, child undernutrition and morbidity: new evidence from India. *Bull World Health Organ*. 83: 210-216.

Pradhan, S, Sharma K. 2011. Nutritional status of Bhil tribal children in Madhya Pradesh, India: a cross sectional study. *Stud Tribes Tribals*, 9: 37-40.

Rao, KM, Balakrishna N, Laxmaiah A, Venkaiah K, Brahmam GNV. 2006. Diet and nutritional status of adolescent tribal population in nine states of India. *Asia Pac J Clin Nutr*. 15:64-71.

Sari, M, de Pee S, Bloem MW, Sun K, Thorne-Lyman AL, Moench-Pfanner R, Akhter N, Kraemer K, Semba RD. 2010. Higher household expenditure on animal- source and nongrain foods lowers the risk of stunting among children 0-59 months old in Indonesia: Implications of rising food prices. *J Nutr.* 140: 195s-200s.

Singh, J, Mondal N. 2013. Assessment of nutritional status: a case of tribal children in Assam, Northeast India. *J Nepal Paediatr Soc*. 33:1-7.

Walingo, MK, Kidake FM. 2013. The influence of household procurement strategies on food intake and nutritional status of pre-school children in rural Western Kenya. *Sustainable Agriculture Research*. 2:109-115.

World Health Organisation (WHO). 1983. Measuring change in nutritional status. World Health Organisation, Geneva.

World Health Organisation (WHO). 1995. Physical status: the use and interpretation of anthropometry: Technical Report Series no. 854. Geneva.

www.unicef.orgindianutrition.html, accessed on 15.07.2013

Xaxa, V. 2011. Children of India: rights and opportunities. IHD - UNICEF Working Paper Series. UNICEF. New Delhi, India.

15

Nutritional and Haematological Status of Migrant Pano Community of Slums in Bhubaneswar, Odisha, India

—*Sunil Kumar Gouda*

Abstract

The present paper highlights the nutritional status and haematological profile of Pano- a Schedule Caste inhabiting in slums of Bhubaneswar. For present investigation 24 hours dietary recall method was used to assess the dietary intakes of children. Anthropometric measurement such as height and weight were recorded and children were classified according to WHO criterion (Z-score) using nutritional indices per day per unit consumption of energy and protein value was calculated and the value was compared with Recommended Dietary Allowances (RDA) of ICMR. The study is based upon 111 households comprising 502 Individuals out of which 239 were males and 263 were females. Percentage of subjects suffering from Grade I (35 per cent), Grade II (30 per cent), Grade III (17.8 per cent) and Grade IV (7.8 per cent) malnutrition indicates higher incidence of malnutrition among the studied community. Half of the population belongs to low calorie range. Only 30.63 per cent households consume recommended calorie per day. Study reveals that 13 per cent have low WBC counts, 26 per cent that of low RBC, 29 per cent have low Platelet counts, 23 per cent have low haemoglobin level. The study reveals that the studied population is nutritionally vulnerable.

Key Words: *Malnutrition, Nutritional deficiency, anaemia, Body Mass Index, Anthropometry, Recommended Dietary Allowances, Haemoglobin.*

Introduction

Nutrition is the sum total of the process by which the organism receives and utilises the materials necessary for the maintenance of its function and the growth and renewal of its components. It includes everything involved from the moment food is taken to its use in various functions of the body. Nutritional status is the balance between the intake of nutrients by an organism and the expenditure of these in the process of growth, reproduction and maintenance of body. Nutrients are the components of food that are needed by the body in adequate amounts in order to grow, reproduce and lead a normal healthy life. Nutrition is the cornerstone of socio-economic development and nutritional problems are not just medical problems but "multi factorial" with roots in other sectors of development such as anthropology, education, demography, agriculture and rural development. It has become apparent that lasting improvement in the health and nutritional status of people can be brought about only through a successful attack on the basic problems of poverty and injustice. The old concept that the health sector alone is responsible for all nutritional ills of the community has faded away. It is now realised that a broad inter-sectorial and integrated approach of sectors of development is needed to tackle today's nutritional problems. Freedom from hunger is a basic human right. Karl Marx once appropriately opined that, "Man must eat to live." Among all basic needs, food deserves the highest priority, as it is pre-requisite to human survival. Nutrition is a state of health conditioned by the ecology of food, and of all factors influencing the health status of individual members of the society; nutrition exerts the greatest influence. Therefore, good nutrition is fundamental in its own right and being a sustainable force for health and development, it indirectly contributes significantly towards the attainment of two key goals such as growth in productivity and distribution of benefits stemming from it (Pelletier, 1994). In the global campaign of "health for all", promotion of proper nutrition is one of the eight elements of primary healthcare (WHO/ UNICEF, 1978). Nutritional indicators (WHO, 1981) have been developed to monitor health for all. Greater emphasis is now placed on integrating nutrition into primary healthcare system whenever possible and formulation of "national dietary goals" to promote health and nutritional status of families and communities.

It has been realised that there is a wide gap between people's need and achievement in the area of health and nutrition. By any comparison the health status of the underprivileged section of the society such as scheduled tribes, scheduled castes, landless labours etc is no way satisfactory. The wide spread poverty, illiteracy, malnutrition, absence of safe drinking water and sanitary living conditions, poor maternal and child health services have been traced out as possible contributing factors for deplorable health conditions and nutritional status prevailing among these vulnerable populations. Malnutrition poses the greatest challenge to the health of India's people and to India's development. The worst victims of perpetual hunger and malnutrition are the vulnerable sections of poor underprivileged communities of India. Assessment of nutritional status of people is one of the national goals.

Hence present study is an endeavour to assess the nutritional status of the migrant Pano (Scheduled caste) communities of Slums in Bhubaneswar, the incidence of Protein Energy Malnutrition and over nutrition among the adults and pre-school children, nutritional health and haematological profile of the population as an indicator of nutritional health.

Relevance of Anthropological Perspectives in Nutritional Studies

The nutritional status of a population is more dependent on the quality and quantity of food consumed by its individual members. Nutrition must be concerned with social, economic, cultural and psychological implication of food and eating. Nutrition as a biological science seems to have marvelled in achieving one of its major historic goals: the derivation of nutrients content in variety of foods, their nutritive values and information on the amount of each nutrient essential for human beings at different ages to maintain optimal nutritional health. This accomplishment has provided a pragmatic basis for planning nutritious and well balanced diet in terms of diets regularly consumed by the community members.

Experience gained from nutritional interventional programme in both industrialized and non-industrialized societies, however, has shown that it has been possible to determine whether diets in common use by different segments of the community are adequate or inadequate to meet their nutritional requirement. There is no denying the fact that diet of people of a given society apart from consideration of economics, agricultural production and disease is greatly influenced by the cultural values placed on food stuffs that may be unrelated to their nutritional values. Therefore, understanding of nutritive values of different food items would help the nutritional planner to determine which foods are not consumed in enough amount by the population and whether the intake nutrients is qualitatively appropriate or not. However, if the culturally prescribed diet of the indigenous population is nutritionally balanced and meeting the nutritional requirements of its members, there should not be any conscious effort on the part of the nutritional planner to recommend diets of other cultures substituting the indigenous and culturally compatible diets in the name of nutritional planning. It is the *opinion of many* ***anthropologists*** that we should make all efforts to reinvigorate studying the science of culturally valued nutritional health and helps provide alternatives to solving the problems of nutritional health at community level. On the other hand, numerous illustrations cited by the nutritionists are replete with evidences with regard to negative impact of cultural tradition on nutritional status. For instance, total elimination of Vitamin A deficiency from the populations of Indonesia and Thailand failed in early sixties through the supplementation of green leafy vegetables because there was cultural prejudice against eating some of the green leafy vegetables. Nevertheless, green vegetables were introduced in to the diet of Zulu population to eliminate Vit-A deficiency because there was no cultural inhibition to consume green leafy vegetables; eggs, on the other hand, were not accepted on religious ground.

From seventies onwards, at a global level, nutritional scientists became interested in understanding the problems of hunger and malnutrition from multidisciplinary perspectives. This has led to the involvement of social scientists including anthropologists in nutritional planning and action programme. However, The biological anthropology of food and nutrition emphasizes the study of nutrition related physiological process with environmental and population characteristics, the differential effects of foods and nutrients on physical growth and development in populations belonging to different socio-economic and socio-cultural categories, and the processes of bio-cultural adaptation of human groups to nutritional stress in different eco-cultural regions. The social and cultural anthropology of food mostly focus on the social-institutional arrangements and the detailed cultural practices and beliefs associated with the food ways, food habits, and food ideology of different communities inhabiting varied ecosystems. The impact of ecology on the food system and food culture of any society is no less substantial since food is a product of cultural ecology. More often than not, the natural resources both biotic and abiotic elements provide wide range of choices, opportunities and alternatives to people to select and grow varieties of crops to meet their food needs. For example, the geo-ecology of an area may suggest the types of crops to be cultivated but what type of crop is to be produced, how it is to be processed, prepared, consumed, shared, stored, distributed including the technologies and resources to be utilized, are by and large determined by the people and their food culture.

The People

The Pano community belongs to scheduled caste category and represent the Indo-Dravidian racial stock, belong to Hindu religion. But now-a-days some of them have converted to Christianity as they are considered as shudra or untouchable by upper caste Hindu. They believe that better living condition, education, better life style etc. can be achieved if they adopt Christianity. Beyond that they will be rid off from social stigma of shudra and untouchablity. The members of Pano caste are mostly migrated population from Kandhamal District, and settled in different slums of the city, they migrated from their native place in search of better livelihood. Most of these people are engaged as manual labourers or wage labourers, some of them are engaged in various private sectors as manual labourers, peon, daily wage labourers etc., truck and taxi drivers, sweepers. Nearly 70 per cent of adult male and female members of the family are seen engaged as part time household workers. They constitute a section of multi-ethnic community. Majority of households of this community have single living room, with no specific kitchen room or toilet. They perform their cooking inside the living room though some have outside kitchen. They have no latrine, therefore they have open defecation practice. The food habit of the Pano is both vegetarian and non-vegetarian. However, due to their low income, they rarely go for non-vegetarian diet. Majority of people have smoking and drinking habits. As they engage themselves in daily wage labour activity, drinking country liquor is a regular feature of the Pano. Both male and female have the practice of chewing Tobacco, Gutka, etc. Majority of these people depend upon open well water for drinking and other activities. There is no proper drainage system and latrine in the locality so the ambience is full of garbage i.e. waste products and human and animal droppings. Majority of Pano depend upon nearby Government dispensary and capital hospital of the city for their health problems. The children of the community go to different Anganwadi Centres and nearby Government Primary School, although high incidence of dropout can be witnessed. The dropout rate is higher among girl-children as they are forced to help their mothers in household activities.

Materials and Methods

Intensive fieldwork is the most important component of anthropological study. Field work is accomplished through application of different techniques. The present study was carried out in different slum areas where the Pano families' reside. The study is based upon 111 households comprising of 502 populations, out of which 239 are males and 263 are females.

For collection of data, different techniques were adopted. Interview methods for Diet history survey, schedule method (Household census schedule) for their food habits, health and Sanitation, addictive habits, health facilities etc.; Observation method for different clinical signs of nutritional deficiency. For the study of haematological profiles like-RBC, WBC, Platelets, Haemoglobin a semi-automatic cell analyser (sysmex-21k, Japan) was used. Intervenous blood samples were drawn irrespective of age, sex, occupation, income. Even blood sample were collected from pregnant women, with their proper consent. Immediately after collection of blood from the syringe, it was poured into the anti-coagulant EDTA-3K tubes and packed. All the samples were brought to the laboratory in freezing condition with proper care and attention.

Different statistical methods were adopted for analysis of data. Mean and standard deviations of all data are taken. Comparisons were also made with different groups and parameters to notice difference. The anthropometric measurements were taken by using standard techniques of Weiner and Lourie (1981). Body mass index (BMI) was computed as weight (kg)/height (m^2). Nutritional status was evaluated on the basis of internationally accepted cut off points of BMI.

Results

The Age Sex-wise distribution of sample is presented in Table 15.1. It is clear that the prevalence of females (52.39 per cent) is greater than males. In all age group, the number of females is greater than male except in the age group- 15-20, 25-30, and 30-40. More number of male (17.15 per cent) and female (16.73 per cent) were found in age group 5-10. Occupation range of the community reveals that out of 502 individuals, only 181 individuals are engaged in different occupation.

Table 15.1: Distribution of Population according to Age and Sex

Age Group	*Male*		*Female*		*Total*	
	No.	*per cent*	*No.*	*per cent*	*No.*	*per cent*
0-1	6	2.51	11	4.18	17	3.38
1-2	5	2.09	11	4.18	16	3.18
3-4	12	5.02	20	7.60	32	6.37
5-10	41	17.15	44	16.73	85	16.93
10-15	22	9.20	36	13.68	58	11.55
15-20	24	10.04	22	8.36	46	9.16
20-25	24	1004	31	11.78	55	10.95
25-30	31	12.97	24	9.12	55	10.95
30-40	41	17.15	32	12.16	73	14.54
40-50	19	7.94	14	5.32	33	6.57
50+	14	5.85	18	6.84	32	6.37
Total	**239**	**100**	**263**	**100**	**502**	**100**

More males are engaged than females in different occupation. Majority of them engaged in manuals works. Both male and female are engaged in different type of occupations. Majority of occupant (male and female) are engaged in manual works. Almost equal proportion of male and female are engaged as daily wage labourer. Only 1.86 per cent of males and 4.05 per cent females are engaged in agriculture & related activity. No one engaged in goverment jobs.

Income range of the community reveals that majority of the individuals belongs to low income range i.e. Rs 3000 Rs 5000. Only 4.62 per cent of males and 3.17 per cent of females belong to income range >Rs7000. The prevalence of males and females decreases as the income range increases.

Table 15.2: Distribution of Population according to Occupation

Occupation	Total Population	Male		Females	
		No.	Per cent	No.	Per cent
1. Agriculture	05	02	1.86	03	4.05
2. Manual Works (Daily Labourer)	132	78	72.89	54	72.97
3. Skilled	13	08	7.47	05	6.75
4. Business	06	04	3.73	02	2.70
5. Private Jobs	25	15	14.01	10	13.51
6. Govt. Jobs	0	0	0	0	0
	181	**107**	**100**	**74**	**100**

Out of 111 households consisting 502 individuals, only 181 individuals are engaged in different occupation. More males are engaged than females in different occupation. Majority of them engaged in manuals works.

Majority of individual have attained education < 10th class, out of these majority are of young children. In illiterate group, the females have greater prevalence (24.06 per cent) than that of males (13.98 per cent). More females (36.46 per cent) were merely primary educated than that of male (26.27 per cent). Greater incidence of male (41.52, 18.22 per cent) than that of female is observed in literate <10th and >10th class.

Table 15.3: Distribution of Population according to their Literacy Status

Literacy Status	Male		Female		Total	
	No.	Per cent	No.	Per cent	No.	Per cent
Illiterate	33	13.98	64	24.06	97	19.32
Literate (Primary)	62	26.27	97	36.46	159	31.67
<10th class	98	41.52	72	27.06	170	33.86
>10th class	43	18.22	33	12.40	76	15.13

Addictive habits of among Pano show that greater incidences of male than that of female are addicted to tobaccoism or alcoholism or both. Higher incidence of smoking is found in male (94.52 per cent) than female. Similarly, alcoholic consumption is higher among male (96.1 per cent) than female. Tobacco chewing habits (39.44) is more than smoking and alcoholic in case of female; although males (60.55 per cent) are more tobacco chewers than females.

Table 15.4: Distribution of per day per unit consumption of calorie among the Pano

Calorie Range (in K.cal)	*No. of Households*	*Per cent of Households*	*Mean Calorie Consumption*	*ICMR Recommended value*
<1800	17	15.31	2127 kcal	2875 Kcal/day
1800-2000	12	10.81		
2000-2200	21	18.91		
2200-2400	34	30.63		
2400-2600	19	17.11		
2600+	8	7.20		

Table 15.5: Distribution of Per day per unit consumption of Proteins among the Pano

Protein Range (in gm)	*No. of Households*	*Per cent of Households*	*Mean Protein Consumption*	*ICMR Recommended value*
20-30	19	17.11	40 gm	55gm per day
30-40	23	20.72		
40-50	31	27.92		
50-60	28	25.22		
60-70	10	9		

Table 15.6: Incidence of PEM (Protein Energy Malnutrition) among the pre-school

Sex of respondent	*Normal*		*1st Grade*		*2nd Grade*		*3rd Grade*		*4th Grade*		*Per cent of Malnutrition*
	No.	*Per cent*	*No.*	*Per cent*	*No.*	*Per cent*	*No.*	*Per cent*	*No.*	*Per cent*	
Male	2	8.69	8	34.78	7	30.43	4	17.39	2	8.69	91.31
Female	5	11.90	15	35.71	13	30.95	6	14.28	3	7.14	88.10

Observations made from Calorie Consumption (Table 15.4) reveals that 30.63 per cent households consume 2200-2400 Kcal/day, 15.31 per cent of households consume <1800 Kcal/day and 7.20 per cent households consume above 2600 Kcal/day. Mean average calorie consumption of Pano is 2127 Kcal/day, which is below the ICMR recommended value of 2875 Kcal/day. Information on protein consumption reveals that 27.92 per cent households consume 40-50 gm protein per day; whereas 17.11 per cent of households consume very low protein, i.e. 20-30 gm per day. Only nine per cent of households consume above 60 gm. Protein per day i.e. as per Recommended Protein consumption. Mean average protein consumption of Pano is 40 gm per day, which is also lower than the ICMR recommended value of 55 gm. Similarly, Protein Energy Malnutrition among the pre-school children shows that- incidence of malnutrition is higher in male child (91.31 per cent) than female.

Table 15.7: Distribution of individuals belonging to different calorie range according to their BMI (adults)

Calorie Range (In K.cal.)	*BMI (Kg/m²)*				*Total*	
	<18.5	*18.5-25*	*25-30*	*>30*	*N*	*Per cent*
1500-1800						
M	10	34	2	-	46	19.24
F	4	24	2	-	30	12.55
Total	14	58	4	-	76	31.79
1800-2000						
M	4	26	2	-	32	13.38
F	2	18	2	-	22	9.2
Total	6	44	4	-	54	22.59
2000-2200						
M	2	16	-	-	18	7.53
F	1	10	-	-	11	4.60
Total	3	26	-	-	29	12.13
2200-2400						
M	-	26	-	-	26	10.87
F	-	8	-	-	8	3.34
Total	-	34	-	-	34	14.22
2400-2600						
M	-	24	4	-	28	11.71
F	-	18	0	-	18	7.53
Total	-	42	4	-	46	19.24
Total Male	16	126	8	-	150	62.76
Per cent	6.69	53.38	3.38	-	-	
Total Female	7	78	4	-	89	37.23
Per cent	2.92	33.05	1.69	-		
Total Individuals	23	204	12	-	239	100
Per cent	9.62	86.44	5.06	-	100	

Approximately 35 per cent, male & female were have 1st grade of malnutrition, 30 per cent were have 2nd grade, 17.31 per cent fall in 3rd grade and 7-8 per cent fall under 4th grade of malnutrition. Only 8.69 per cent of male child& 11.90 per cent female child were found normal.

Table 8. Distribution of individuals belonging to different Protein range according to their BMI (adults)

Protein Range (in gm)	*BMI <18.5*	*18.5-25*	*25-30*	*>30*	*Total*	*Per cent*
20-30						
M	13	9	-	-	22	19.20
F	6	8	-	-	14	5.85
Total	19	17	-	-	36	15.06
30-40						
M	16	26	4	-	46	19.24
F	4	21	0	-	25	10.46
Total	20	47	4	-	71	29.70
40-50						
M	-	30	2	-	32	13.38
F	-	32	0	-	32	13.38
Total	-	62	2	-	64	26.77
50-60						
M	-	18	2	-	20	8.36
F	-	14	0	-	14	5.85
Total	-	32	2	-	34	14.22
60-70						
M	-	22	4	-	26	10.87
F	-	8	0	-	8	3.34
Total	-	30	4	-	34	14.22
Total Male	29	105	12	-	146	61.08
Per cent	12.12	43.93	5.02	-	61.08	-
Total Female	10	83	0	-	93	38.91
Per cent	4.18	34.72	0	-	38.91	-
Total Individuals	39	188	12	-	239	100
Per cent	16.31	78.66	5.02	-	100	-

Calorie range to BMI shows that more than half of the adults belong to low calorie range i.e. from 1500-2000 Kcal/day for BMI. It is clear from the Table that approximately 55 per cent of adult individual falls under low calorie range. Only 3.38 per cent of adults have normal BMI. Protein range to BMI reveals that approximately 45 per cent of the adult individuals belong to low protein range for BMI. It should be noticed that 73 per cent of total working individuals are wage labourers and belong to very low income range. Due to less consumption, their protein range to BMI is also very low.

Above all these parameters were measured and analysed to assess the Nutritional status of Pano community.

Analysis of Haematological Parameters

Analysis of different haematological parameters of Pano people reveals that- WBC graph for age group shows approximately 13 per cent of individuals represent below normal WBC count and 3 per cent of that represent above normal WBC counts. Age group 15-20, 30-40, 40-50, 50+ shows more abnormalities. No children were found to have such abnormality. Platelet (PLT) graph for age group shows approximately 29 per cent of Individuals belong to below normal of Platelet count and 1.42 per cent of that belongs to above normal.

RBC graph for age group shows approximately 26 per cent of individuals belong to below normal RBC count and 5.68 per cent belong to above normal count. 20-30 and 30-40 both age groups have equal and more, they do not have abnormality. Similarly, 23 per cent of individuals have low haemoglobin, whereas 1.42 per cent has normal level of haemoglobin.

Overall low WBC, RBC, Haemoglobin and Platelet count is observed among low income group, increasing age group, females and among daily wage labourers. WBCs, an important part of immune system and body's natural weapon to fight against bacteria, viruses and other germs. Due to low WBC count their immune suppressed, and they are more prone to infections. Low RBC and Haemoglobin among low income group and female indicates different types of anamia. Anamia can be caused due to nutritional deficiencies, especially iron and vitamin deficiency which is a regular feature of Pano slum dwellers due to their low RBC and haemoglobin count as revealed from their Haematological studies.

Conclusion

With rapid urbanisation 20 per cent of urban population is forced to reside in slums. Slums are the physical and social expression of inequalities in the distribution of the beneficiaries of economic growth as well as the structure performances and special pattern of the urban economy. Non-availability of basic human necessities in urban slums adversely affects the growth and nutritional status of the slum dwellers. Being the most vulnerable segment of the population under five children, pregnant mothers are at greater risk of undernutrition because of poor purchasing power, insufficient dietary intake and high load of infections.

In the present study, different parameters determining the nutritional health status of Pano community were measured. After analysing all the parameters, it is revealed that this community is more vulnerable nutritionally. As majority of people are wage labourers who belong to low income group, they remain below poverty line. They are unable to assess required nutrients for maintaining proper health condition. They do not get employment every day. They do not have proper access to safe drinking water, health and educational facilities. Very often majority of poor class suffer from diseases like malaria and nutritional anamia, cold and cough, joint pain, scabies, jaundice, asthma due to intake of improper nutrients and hygienic living conditions. Some females are also suffer from menstrual disorders pain as they remain always in unhygienic condition, and that may be due to hormonal imbalance and excess stress. People of this community always remain in stress and strain due to lack of adequate work. To get better earning, for better living standards, they have migrated from their native place, where there is very little scope for livelihood. More than half of people of this community migrated due to pre and post-communal riot at Kandhamal, their native district. Though government has implemented various programmes for the improvement of health, education, eradication of poverty, combating various communicable and non-communicable diseases, employment guarantee schemes, food security act for the overall improvement of weaker sections, but most of such programmes are swallowed by wide spread corruption and hardly reached to needy clientale for whom it is meant. The present study throws light on the nutritional and health vulnerability of Pano community and draws attention of researchers, different voluntary organisation, NGOs as well as government agencies for overall development of Pano people.

References

Ali, A 1992. Nutrition. In: Alok Mukhopadhyay (Ed.): State of India's Health. New Delhi: VHAI, pp.1-50.

Basu, SK, Jindal Anil, Gautam K 1990. Genetic and Sociocultural Determinants of Tribal Health. ICMRs Final Report, New Delhi: NIHFW. GOI, Ministry of Home Affairs. 1985. Report of the Working Groups for Development of Tribal during Seventh Five Year Plan: 1985-90. New Delhi: Ministry of Home Affairs.

Balgir, RS, Kerketta AS, Murmu B 2002. Clinical assessment of health and nutritional status of Gond children in Kalahandi district of Orissa. *Ind J Nutr Dietet*, 39: 31-37.

Bailey, KV, Ferro-Luzzi. Use of body mass index of adults in assessing individual and community nutritional status. *Bull WHO* 2005; 73: 673-80.

Bhardwaj, S, Kapoor S. Nutritional anthropometry and health status: a study among Dhanka tribals of Rajasthan. *Anthropologist* 2007; 9(3): 211-4.

Bose, K, Biswas S, Bisai S, Ganguli S, Khatun A, Mukhopadhyay A, Bhadra M 2007. Stunting, underweight and wasting among Integrated Child Development Services (ICDS) scheme children aged 3- 5 years of Chapra, Nadia District, West Bengal, India. *Matern Child Nutr*, 3: 216-221.

Bose, K, Chakraborty F, Bonsai S, Khatun A, Bauri H 2006. Body mass index and nutritional status of adult Savar tribals of Keonjhar District, Orissa, India. *Asia Pac J Pub Health*, 18: 3-7.

Chakraborty, R, Boss K. Anthropometric characteristics and nutritional status of adult Oraon men of Gumla District, Jharkhand, India. *Int J Biol Anthropol* 2009; 2: 1-10.

Chatterjee, P 2007. Child malnutrition rises in India despite economic boom. *Lancet*, 369: 1417-1418.

Chowdhury, SD, Chakraborty T, Ghosh T 2008. Prevalence of under nutrition in Santal children of Puruliya District, West Bengal. *Indian Pediatr*, 45: 43-46.

Deka, U and Das P.K. (1987): The nutrition and physical growth in tribal, Orissa.

Das, DK, Biswas R 2005. Nutritional status of adolescent girls in a rural area of North 24 Parganas district, West Bengal. *Ind J Public Health*, 49: 18-21.

Das, DK, Biswas R 2005. Nutritional status of adolescent girls in a rural area of North 24 Parganas district, West Bengal. *Indian J Public Health*, 49(1): 18-21.

Draper, H.H.: Biological, cultural and social determinant of Nutritional Status Diet and Nutritional status of population and prevalence of Hypertension among adults in rural areas. NNMB Technical Report No: 24, NNMB, NIN, ICMR, Hyderabad-2006.

Ghosh, R, Das K, Bhatia P. Health and nutritional status of Ho preschool children of Orissa. *J Hum Ecol* 2001; 12: 109-13.

Gopalan, C, Sastri BP, Balasubramanian SC 2001. Nutritive Value of Indian Foods. Hyderabad: National Institute of Nutrition (ICMR).

Gopalan, C 1971. Nutritional Atlas of India. New Delhi: ICMR. 1990. Report on Nutrient Requirements and Recommended Dietary Allowances of Indian. Hyderabad: NIN.

Grover, K, Singh I 2006. An analysis of food intake of rural preschool children in Punjab. *J Hum Ecol*, 20: 161-164.

Jai Prabhakar, SC, Gangadhar MR 2009. Nutritional status of Jenukuruba tribal children in Mysore District, Karnataka. *Anthropologist*, 11: 83-88.

James, LF 1998. India-Sector Review of Nutrition Programmes. A background paper prepared for the World Bank, New Delhi.

Kapoor, AK. Tribes of Andaman and Nicobar Islands: then and now. *Dialogue* quarterly 2003; 4(4): 59-68.

Kapoor, AK. Tyagi R, Kapoor S. Nutritional status and cardio respiratory functions among adult Raji males, a hunter and gatherer tribe of the Indian Himalayas. Anthropol Sci 2009; 117(1): 1-7.

Kapoor, S, Kapoor AK, Tyagi R. Socio-Biological Ageing: an inevitable phenomenon – Help Age. *India Res Dev*. J 2000; 6: 5-10.

Kumar, Meta, Sharma Sushma 1988. Measuring Malnutrition – Tools and Approaches. *Journal of Swasth Hind*, 2(9): 17 - 21.

K. Sreenath Reddy, Bela Shah, Cherian Vergese and Anbumani Ramadoss. Responding to the threat of chronic diseases in India, the *Lancet*, October 2005, published online DOI; 10.1016/50140-6736(05) 67341-2

Malhotra, A, Passi SJ 2007. Diet quality and nutritional status of rural adolescent girl beneficiaries of ICDS in north India. *Asia Pas J Clin Nutr*, 16 suppl 1: 8-16.

Medhi, GK, Hazarika NC, Mahanta J 2007. Nutritional status of adolescents among tea garden workers. *Ind J Pediatr*, 74: 343-347.

Mittal, PC, Srivastava S 2006. Diet, nutritional status and food related traditions of Oraon tribes of New Mal (West Bengal), India. *Rural Remote Health*, 6: 385.

Mitra, M, Kumar PV, Chakrabarty S, Bharati P 2007. Nutritional status of Kamar tribal children in Chhattishgarh. *Indian J Pediatr*, 74: 381-384.

Modak B. and Das P.K.: Nutritional status of Bonda Highlanders of Orissa, p. 125-175, in Edited Book "Sustainable development in Bonda Hills" by P. Mohapatra, 2009, Centre of Analytical tribal studies(COATS), koraput.

Nandy, S, Irving M, Gordon D, Subramanian SV, Smith GD 2005. Poverty, child undernutrition and morbidity: New evidence from India. *Bull World Organ*, 83: 210-216.

National Institute of Nutrition 2011. Dietary Guidelines for Indians-A Manual. Hyderabad: National Institute of Nutrition.

National Nutrition Monitoring Bureau 2000. Diet and Nutritional Status of Tribal Population Repeat Survey. Hyderabad: National Institute of Nutrition.

National Nutrition Monitoring Bureau. 2000. Diet and Nutritional Status of Tribal Population. Report on first repeat survey. Hyderabad: National Institute of Nutrition.

Park, JE, Park K 2009. Preventive and Social Medicine. Jabalpur: Benarasi Das Bhanot.

Prashant K, Chandan S 2009. Nutritional status of adolescent girls from an urban slum area in South India. *Indian Journal of Paediatrics*, 76: 501-504.

Rao, BD, Busi BR 1999. Growth progression in physical and physiological variables among Poroja tribal boys of Visakhapatnam District, Andhra Pradesh, India. *South Asian Anthropologist*, 20: 35-42.

Rao, KM, Laxmaiah A, Venkaiah K, Brahmam GN 2006a. Diet and nutritional status of adolescent tribal population in nine states of India. *Asia Pac J Clin Nutr*, 15: 64-71.

Rao, VG,Yadav R, Dolla CK,Kumar S,Bhondeley MK,Ukey M 2005. Undernutrition and childhood morbidities among tribal preschool children. *Indian J Med Res*, 122: 43-47.

Rao, KM, Kumar RH, Venkaiah K, Brahman GNV 2006. Nutritional status of Saharia- A primitive tribe of Rajasthan. *J Hum Ecol*, 19: 117-123.

Report of the Expert Group of the Indian Council of Medical Research (1990), Nutrient Requirements and Recommended Dietary Allowances for Indians, ICMR, New Delhi.

Saheb, SY, Rajendra Prasad M 2007. Growth and development of the Lambadi children of Nalgonda. District of Andhra Pradesh: Bio-Cultural perspective. Kolkata: Anthropological Survey of India, (in press).

Sharma, B, Mitra M, Chakrabarty S, Bharati P 2006. Nutritional status of preschool children of Raj Gond- A tribal population in Madhya Pradesh, *India. Mal J Nutr*, 12(2): 147-155.

Shetty, PS, James WP. Body mass index: a measure of chronic energy deficiency in adults. *FAO Food Nutr Pap* 1994; 56: 1-57.

Singh, IP, Bhasin MK 1968. Anthropometry. Delhi: Bharti Bhawan.

Singh, N, Mishra CP 2001. Nutritional status of adolescent girls of a slum community of Varanasi. *Indian J Public Health*, 45(4): 128-134.

Som, S, Pal M, Bhattacharya B, Bharati S, Bharati P 2006. Socio-economic differentials in nutritional status of children in the states of West Bengal and Assam. *J Biosoc Sci*, 38: 625-642.

Tanner, JM 1978. Foetus to Man: Physical Growth from Conception to Maturity. Massachusetts: Harvard University Press.

Tyagi, R, Kapoor S, Kapoor AK. Environmental influences and health status of elderly. *Open Anthropol J* 2008; 1: 14-8.

Thimmayama, BVS 1987. A Handbook of Guidelines in Socio-Economic and Diet Survey. Hyderabad: National Institute of Nutrition.

Uppal, M, Kumari K, Sidhu S 2005. Clinical assessment of health and nutritional status of scheduled caste preschool children of Amritsar. *Anthropologist,* 7: 169-171.

Varsha Z, Rohinidevi, Asha A 2008. Effect of supplementation of bio fortified biscuits on haematological parameters of selected adolescent girls of Marathwada Region. *Ind J Nutr Dietet,* 45: 520-529.

Venkaiah, K, Damayanti K, Nayak MU, Vijayaraghavan K 2002. Diet and nutritional status of rural adolescents in India. *Eur J Clin Nutr,* 56(11): 1119-1125.

Vijaya, Raghavan K, Singh D, Swami Nathan MC 1971. Heights and weights of well-nourished Indian school children. Ind J Med Res, 59: 648-654.

WHO 1983. Measuring Change in Nutritional Status. Geneva: World Health Organisation, pp. 63-101.

World Health Organisation (WHO) 1995. Physical Status: The Use and Interpretation of Anthropometry. WHO Technical Report No. 854. Geneva: WHO.

WHO Diet, nutrition and prevention of chronic disease. Report of a joint WHO/ FAO expert consultation. Technical series no.916. 2003.

WHO 1995. Physical status: The use and Interpretation of Anthropometry, Report of a WHO Expert committee, Technical Report Series, No. 854, Geneva: WHO.

Theme – 6
Food Security and Healthcare

16

Role of Self Help Groups in Providing Food Security and Health Care with Reference to Children: A Study in Andhra Pradesh

—*M. Sreevidya*

Introduction

Rural women, in particular, those belong to the economically and socially weaker sections of society have limited access to resources or employment opportunities that would make them financially independent. They survive in conditions that do not permit them to meet minimum consumption needs. Even the money earned by them from hard physical labour is not controlled by them. Institutional credit is difficult to access as women rarely have any property rights to mortgage or collateral to offer, and they have to depend on rural money-lenders for their production-cum-consumption needs. This entraps them into a vicious cycle of perpetual poverty and indebtedness. There is increasing awareness and consciousness that rural women have been underestimated and discriminated against in all walks of life, despite their ample contribution to the household economy and, in turn, the national economy. The UN Commission on Status for Women observed that 'women, who contribute half of the world's the population virtue of an accident of birth, perform two-thirds of the world's work, receive one-tenth of its income and own less than one-hundredth of its property'. In India, women produce 30 per cent of all food commodities consumed but get only 10 per cent of the income, and own 10 per cent of the property or wealth of the country (Sarmistha Choudhoury, 2004).

Self Help Groups (SHGs) are usually informal groups whose members have a common perception of need and importance towards collective action. These groups not only promote savings and use the pooled resources to meet the emergent needs of their family members, including the consumption needs, they try to facilitate themselves to develop/ empower in some basic fields, like education, occupations child welfare etc. The SHGs were set up by NGOs with the assistance of subsidised loans, bank linkages, to provide individual loans for women beneficiaries and individual/joint liability for repayment. In the beginning stress is less on income generation activities or asset creation and more on production-cum-consumption loans, ensuring quicker repayment and fewer borrowers fatigued in loan repayment. The repayment rate is much higher in SHGs as there is no subsidy involved. The Development of Women and Children in Rural Areas (DWCRA) programme has been an important adjunct to the Integrated Rural Development Programme (IRDP), which was launched in 1980. The IRDP is a unique programme in view of its vast coverage, but, it does not target women specifically. In 1998, women covered under SAPAP credit were about 34 per cent of the total 66,000 households (UNDP, 2000). Rural development means overall development of rural areas to improve the quality of life of rural people –men, women and children. It is integrated process, which includes socio-economic, political and overall development of the poorer sections of the society. Government provide basic minimum facilities to the people of particularly marginalised sections with primary education to all women, men and children, primary health, infrastructure etc.

This chapter describes the socio-economic profile of the woman self help groups in AP. It also examines the efforts being taken by the woman self help groups in providing care in the case of children in the community. Also deal that the SHG women take initiation in all development programmes implemented by the government. Particularly in children's programmes like eradication of child labour, encouraging Bridge Schools, Sarva Siksha Abhayan, Pulse Polio, nutrition and healthcare of pregnant women and children, mid-day meals etc.

Objectives of the Study

The overall objective of the present study is to analysis the participation of women and children related programmes implemented by the government or NGOs through SHGs in Krishna District of Andhra Pradesh. However, more specifically: To study the education and healthcare of the children in the families of the SHG members after joining SHGs.

Area and Sample

The area of the study consists of Krishna district in Andhra Pradesh and its Scheduled Caste populations. The area of the study covers thirty villages from six Mandals in three revenue divisions of Krishna district in Andhra Pradesh. Care was taken to cover the SHGs formed in the year 1999 and before to understand the processes of changes occurred among the members of SHGs. Thus, 6 Mandals, 30 Villages, 90 SHGs and 270 SHG members were selected in Krishna district of Andhra Pradesh for the Study.

Methodology

The data for the study were collected from both primary and secondary sources. The secondary data relating to the distribution of Scheduled Caste populations and woman self help groups, programmes being implemented for the empowerment of women belonging to the Scheduled Castes were collected from the concerned official records. The primary data were collected through anthropological field work using interview schedule, observation, formal and informal interviews and case study method. An interview schedule was prepared and pre-tested to administer among the woman members of the selected SHGs.

SHGs

This idea of formation of groups usually takes place for a common need, but often just one person or an organisation takes initiative. The key elements to form a structure of SHG are:

All the members of the SHG should be the inhabitants of the same village. The age of the SHG members should be between 18 to 60 years and each member must be willing to accept the objectives of the group. Like-minded and interested individuals discuss about this concept and they joined and involve more people. It is small in size and membership of a group is restricted to 10-15 persons.

After formation of the group, the members should select a name by which the group is identified. The group may register itself as an organisation. It is non-political and voluntary and follows democratic culture. Potential members in the community are informed about the concepts, information about the benefits, programmes and activities of the government and NGOs. The group elects leaders democratically and conducts meetings. They discuss the group savings, rotation of group funds, bank loans, repayment of loans, social and community action programmes. SHGs have transparency among themselves and they have collective accountability with regard to financial transparency. Members should not be involved in any criminal and anti-social activities. A SHG formation stage generally takes 1-6 months and should reach its self-reliant within 18 to 24 months.

The activities

To achieve its objectives, the society has formulated the following activities with a view to participation of women related to child welfare programmes particularly in the rural areas:

- Creation of awareness among the member SHGs about issues like children education, child labour, health, sanitation, child marriages and other women-specific problems and motivating them to act collectively;
- Developing linkages with government and non-governmental organisations and taking up welfare activities for the social and economic emancipation of member SHGs and their families;
- Supply of information to the member SHGs from different government and other sources.

The SHG members appear to have experienced that education gives an impetus for occupational and social climbing. A growing number of their organisations are in the name of Dr Ambedkar and most of them define their first task as education for their own community.

Thus, this awareness appears to contribute to the increasing trend of education among them. The sphere of using law for their protection is another factor in enhancing their social position. Instances have been noted where they could use the court of law and police against the dominant and higher castes. The recent trends of organising together against any such kind of suppression are visible.

Health and nutrition initiative

Imparting nutrition education to beneficiaries of selected villages is another important activity of the NGO. Through its PRA study the APRPRP has notice the incidence of malnutrition and iron deficiency among the children.

Spending on health is a productive investment. It can raise incomes, particularly among the poor, and it reduces the toll of human suffering from ill health. Good health, however, is a fundamental goal of development as well as a means of accelerating it. Targeting health as part of development efforts is an effective ways to improve welfare in low income countries. Evidence gathered over the gap between poor and rich countries has been narrowing (Jayarama Reddy, 1995). Apart from this the prevalence of large-scale superstitious beliefs, ignorance, illiteracy and poverty among the people of the area, have not allowed them to think about the importance of nutritious food. Hence, it has set of objective of improving the nutritious level of children of target families.

To create awareness about the nutritious value of food and to improve nutrition levels of beneficiaries APRPRP organised training programmes. The training needs identified for nutrition programmes included: awareness creation on utilisation of Balavadi /Anganwadi centres, preparation of special diet with locally available food stuff for children, pregnant women and lactating mothers and encouraging vegetable cultivation.

As a part of these programmes, the government encouraged the beneficiaries to grow kitchen gardens and cultivate vegetables to make vegetables accessible to the target families. It also supplied milk and special diet to the children and pregnant and lactating mothers in the villages.

National Rural Health Mission

Detail action plan has prepared for conducting training programmes to the field functionaries of Medical and Health Department, ICDS and IKP staff for conducting fixed NH Days in the district. The AP Rural Poverty Reduction Programme has specific strategies to address issues of health and nutrition services that are essential for the rural poor through empowering the communities and creating an enabling environment in project mandals of Andhra Pradesh. The project ensures that there is internal convergence between health, nutrition and disability strategies undertaken within the project so that all these activities play a synergistic role in enhancing the quality of life of the communities. The overall goal of this component is to improve health and nutrition status of poor. The specific activities that will be undertaken under this component will be:

- Build capacities of the communities for improved health and nutrition seeking behaviour and thus reducing the risks.
- Develop a care of health and nutrition activists selected from within the communities.
- Nutrition centres to fill the gaps in the child health and nutrition services in the community.
- Some focused health initiatives under Community Investment Fund (CIF).

Workable models for health / social insurance Pilot Schemes to improve access control over their own health. The project will play a catalytic role in improving the health and nutrition status of poor closely working with the line departments and private sector to improve access to basic and nutrition services to rural poor demonstrate through pilots the options for providing healthcare as well as safety nets to poor.

NMBS (National Maternity Benefit Scheme)

The programme was started in 1995 at national level, and it is one of the national social assistance programmes. Mainly this programme meant for the pregnant women to give assistance before the delivery (8-12 weeks).

ICDS (Integrated Child Development Services)

Services of this scheme were started in 1975. Services are implementing through Anganwadi centres. Pregnant women, lactating mothers, children below the age of six years come under this programme. The main aim of this programme is improving mental and physical growth, reducing mortality rate, ill health and diseases, and reducing the lack of nutritious food. The main aim of this programme is to increase the knowledge of nutritious food among mothers and providing nutritious food, vitamin A, Iron, Folic Acid, Vaccination, drinking water facility and sanitation to the beneficiaries. Currently, Government of India implementing the programme in all States.

Supplementary Nutrition Programme

The state government of Andhra Pradesh implementing the programme by selecting 40 per cent of the pregnant women from SC, ST, BC, labourers come under Below Poverty Line (BPL). The specialty of this programme is to give nutritious food as Ready to Eat-Ret to the children from six months to five years of age group. The SHG women in the study area are above 80 per cent, sending their children to the Anganwadi centres. There the children will get nutritious food like EGGs, fruits and healthy oowder (millets and cereals).

Adolescent Girls Scheme

The adolescent girls, the young women, the mothers to be, represent an age group which has received little attention earlier. But now the government, the national and international organisations are decisive to lay importance for changing the situation of both girl children and women in lower and deprived groups like scheduled castes.

As a girl child, women's role is very limited and restricted to a certain level. Their freedom is controlled and disciplined by the mothers/elders. In the field of work a strong discrepancy is found in the education of female as compared with male child mainly because of lack of financial support and ignorance. In many families, the education of male child is a precedence/priority over the education of female child. This situation often brings frustration and problems among the siblings. However, due to recent change in the attitude towards girl's education, the girls are allowed to have education and orthodox familial restrictions are gradually disappearing.

This programme mainly aims to improve literacy, mental and physical health, develop knowledge of nutritious food and provide nutritious food to adolescent girls from below poverty line, and to give skills and to became ideal mothers in future. Not only that, the adolescent girls had given trainings about the society, to get change in their attitudes, importance of primary education, to give skillfull vocational trainings.

Balika Samvriddhi Yojana (Girls Welfare Scheme)

This programme meant for the welfare of female children and women among below poverty line. This scheme is to provide nutritious food after delivery of women.

Access to Food, Health and Education

The data collected from the field work includes information on food, health and education to understand how far SHG women access these basic needs. The data on the access to food shows that at the time of joining groups, 59 per cent of members had access to three meals a day; after joining the group, now 98 per cent of the members have access to three meals a day. Now, the percentage of households having food for all the household members is high when compared to the percentage of households at the time of joining the group.

Table 16.1: Access to Food, Health, and Education

Sl. No.	*Access to FH&E Particulars*		*Yes*		*No*	
			No	*Per cent*	*No*	*Per cent*
1.	3 Meal a Day	Before	26	1.07	146	6.01
		Present	155	6.38	115	4.73
2.	Increased quality of Food	Before	256	10.53	138	5.68
		Present	262	10.78	8	0.33
3.	Sending Children to Schools	Before	215	8.85	162	6.67
		Present	262	10.78	8	0.33

Of the total 270 members at the time of joining the group's 82 per cent of the members sending their children to the schools.

Data on the average income of the HH 145598 and the objective wealth ranking of the sample member HH, which shows that 45 per cent are in the self sufficient or in the surplus categories, confirms the same.

Food Security: Rice Credit Line

In technologically developed countries, nutritional deficiency diseases have largely been eliminated. Unfortunately, most of the developing countries, including India, malnutrition continues to be a public health problem. Due to poor socio-economic conditions majority of the Indian population are not in a position to afford the least expensive balanced diets, as a result the prevalence of malnutrition is high among the poorer section particularly scheduled castes and tribal communities of population.

To eradicate malnutrition and nutrition-related health problems, Government of Andhra Pradesh implemented a Food Security Programmes at village level to the SHG women. The main objective of this programme of the food security is taken up by the Village Organisation with the objective to ensure the food for poor families in advance. This includes access to the rice taken from the Fair Price Shop under Public Distribution System (PDS).

The Village Organisation (VO) will ensure all the members of the SHGs should be enabled to take rice from fair price shop and direct purchase from Rice Mills every month by providing credit, if needed to members through the respective SHGs.

The VO also discusses the consolidated proposal for all SHGs, in its executive committee meeting held for every month and given approval; of the EC, will procure the indented quantities. The impact of this programme are (a) the poor are assured the food for all the days in a month (b) the quality, quantity and reasonable rice are ensured (c) the intake of food is increased by the poor particularly children.

Participation of SHGs related to Child Welfare Programmes

Parents Committee

Parents committee is a CBO formed at the village level. Village organisations have formed parents committees in all the villages.

There are 700 members in 50 parents committees. It meets once in every month to work in close cooperation with the men and women groups. Identification of child-bonded labourers. Promotion of child education particularly girls, encouraging rejoin schools to the dropouts, preparation of herbal medicine etc, are the main activities of the committees.

Vidya Committee

The discussion with the projects personnel reveals that initiation of educational programmes with focus on minimum literacy levels and awareness to beneficiaries are necessities to make them understand the socio-economic process in the village. As a part of strategy the government has formed Vidya Committees in all the villages. Both elders and youth, who are interested in the promotion of education in the villages, are drawn into this committee. Promotion of adult education, motivating adults for education, eradication of bonded labour practices, initiation time bound literacy development programmes etc. are the important activities of Vidya committees. They function through the adult education centres established in the villages and take steps for community education programmes through slide films and cultural media. Also SHG women participate in cooking food for Children in the Schools in the Programmes of Mid-Day-Meals. Now-a-days women particularly concentrates on personal hygiene of them self as well as children.

ECE (Early Child Education)

In the year 2008-09, ECE centres have been set up under the supervision and governance of community. The children in age group of three to five years are provided with Early Childhood Education in these centres. Community governed, community managed ECE centres have shown significant improvement in physical, mental and cognitive development of these children.

Effective monitoring mechanisms and public reporting of the performance of the students make these centres stand apart. Members of the VOs (Village Organisations) and parents are participating in the ECE activities regularly. The success of the ECE model has created demand in other tribal areas of the State. On similar lines, in 2009-10

SERP proposes to support MMS (Mandal Mahila Samakhyas) in five districts who have come forward to set up and govern ECE centres in their respective mandals. The SHG women will be empowered to identify the villages and the children between the age group of three to five years and make necessary logistic arrangements for setting up of ECE centres. The enrolment in Primary schools in the tribal areas has improved significantly in the last four to five years.

Summary

The project will play a catalytic role in improving the health and nutrition status of poor, closely working with the departments and private sector to improve access to basic and nutrition services to rural poor demonstrate through pilots the options for providing healthcare as well as safety nets to poor.

Under Social and Human Development (addressing Millennium Goals) Social Action, for gender equity, recognising the people with disabilities among the poorest of the poor. Health and Nutrition interventions quality education through community participation as intergenerational poverty alleviation strategy were covered. And some of the major findings of the study are:

1. Members of the groups had access to three meals a day after joining the SHG.
2. Reduction in infant mortality.
3. Reduction of Deaths of children below five years of age.
4. Reduction of disability incidences due to malnutrition.
5. Improved hygiene and sanitation and control of spread of communicable diseases.
6. Supply of safe drinking water.
7. Decreased child labour and dropout from schools.

Gender programmes helps preventing discrimination against girl child, human trafficking, domestic violence helps women to increase their understanding of intra-family equity issues, decisions making levels, free mobility and necessity of building safe environment.

References

Gopalan, C. Nutrition (1989), Health and National Development. Chap.4 Under Nutrition measurement and Implication. spl.pub. Series No. 4 Nutrition Foundation of India.

IKP (2010): Indira Kranthi Padham Progress Report.

Jayarama Reddy, P (1995): Health in Developing Countries Success and Challenges, Delta Publising House, Hyderabad.

NABARD (2010): Status of Micro-Finance Report 2009-10.

Sreevidya, M (2012): A thesis submitter to SV University "Micro Finance and Changing Livelihoods: A Study on Socio-Cultural Change among Scheduled Castes in Krishna District of Andhra Pradesh."

Index